Simple Biology Investigations

Solutions for doing science in the classroom.

Christopher P. Garside

Seven Sides Publishing

Seven Sides Publishing of Cypress, TX, has a mission to improve teaching and the understanding of science. To contact us, send an email to simpleinvestigations@sevensidespublishing.com or visit us at sevensidespublishing.com.

ISBN: 9798690731878

Published by: Seven Sides Publishing, Cypress, TX.

Table of Contents

Introduction

To help teachers teach science through investigations, Seven Sides Publishing has provided a series of lab manuals for Elementary Science, Middle School Science, Physics, Chemistry, Biology, Environmental Systems, and Earth & Space Science. These manuals are a rich resource for structure and investigations. I have noticed a shortage of user-friendly labs that easily allow teachers and students to perform investigations in a timely manner. Too many labs have too much busy writing where teachers do not want to take the time to read everything to figure out if it would be good for them to use with their students. If the teachers do not want to read it, do you think the students do? So I have taken a lot of the traditional labs that have been around for decades and simplified them so they are easy to read and perform. I have also made and added some new original labs that have never been seen before. There has been an effort to try to have teachers do more investigations with their students, but there is no plan or solution to deal with the real issues teachers have in preparing to do this. The book How to Teach Science Through Investigations has the plan, and the Simple Investigations Lab manuals have the solutions so students can learn science through investigations with minimal effort. This series will make your classrooms more efficient, where students learn content and practice skills simultaneously. Science is a process of doing. Doing this process is the most important way for students to learn science and be able to use it in the future. We live in a culture where science-literate people are needed for jobs, but too few can be found. If you incorporate these labs with virtual labs (that I will point you to in each section of the lab manual), skill/math practice, and concept maps, you will not need to fill in gaps by giving lectures. All content can be learned through investigations and practice. Remember, we only remember 5-20% of what we hear. That 20% is when you are interested in the content. But hearing practices no science process skills and does not activate any higher cognitive thought. Lecturing is not a good option. We remember 75-80% of what we do/experience and 90-95% of what we teach. Investigations allow us to keep our students in these higher retention percentages. The main reason this works is that students spend more time in class at higher levels of Bloom's Taxonomy, staying in zones C and D of the Rigor Relevance Chart when they perform investigations. And if you add the physical way they are stimulated with the hands-on experience, you cannot deny the level of learning will be much higher while students perform investigations. This manual gives you the resources you need to teach Biology through investigations.

We separated each of these sections in the manual like you may separate your units in the class. There have been many studies on how best to present the order of content in friendly ways. This scope and sequence will follow a narrative of the Story of Life. It will start with biochemistry and the characteristics of life in the first unit. Then we give a foundation for what forms life; the interactions in ecology and evolution. Each unit after will show how life evolved from bacteria to protists to multicellular life, making complicated plants, fungi, and animals. I will include concept maps at the front of each section that shows the vocabulary and visual clues to how concepts relate to each other. These concept maps are a great way to organize information and talk to the students for them to see how ideas work together and chunk information at higher cognitive levels. At the beginning of each lab, we put the materials you will need in boldface in the directions. This design saves time and space to help with your preparation. There is also a safety question in boldface just after that for you and your students to evaluate. It says," Looking at the material and lab we will be using, what are the safety precautions we should take to protect ourselves and materials during this investigation." Make sure to read through the lab to help you better answer this question.

Virtual Labs

Hands-on labs are not the only way for students to learn science, but they are the most effective. However, many virtual labs should be used with these hands-on labs. Many investigations physically cannot be done hands-on, so some experiments will have to be done virtually. There are two sources that I have used in the past that have resources for Biology. **PhET.colorado.edu** is free to everyone and is great to use. **PhET.colorado.edu** has various Biology activities that you can explore to go through their simulations. They are also easy to download and print. **ExploreLearning.com** is expensive, but the quality and quantity of their products are great. When you click on a Gizmo, you can also click on lessons and find the Student Explorations that go with each Gizmo that you can modify, download, or print. They are written at a very high quality, making the students think like a scientist. At the end of each section of this lab manual, I will include a list of virtual labs from these organizations that would be great to use with these labs. Please remember virtual labs should never replace hands-on labs. If the students can learn the content live, that should be the priority because it is more of an experience that will be remembered. There are many other virtual simulations out there, but none so far have moved me to use them over the two I have mentioned here.

Probe-ware

This lab manual has lots of labs that use probe-ware. Students must learn how to use probe-ware; this means teachers need to learn how to use probe-ware. Many companies use digital probe-ware with all the research, development, testing, and forensics they do. Having this skill can give career opportunities and help students become more marketable for jobs if they are familiar with using probe-ware. Hooking everything up is just as easy as charging your phone. When I was a High School Science Technology Coach and researched which companies and devices would be the most user-friendly to students, I found using Vernier Probe-ware was better for high school students, but PASCO seemed better for middle school students. Both are giants in the probe-ware industry for education. Since this Lab manual and the series were written with High School in mind (many of these labs can be used for middle school classes because they are so simple) and I am more familiar with Vernier, I will be referring to Vernier Probe-ware. However, PASCO would be a great alternative.

Interfaces are devices that the probes are connected to that talk with the program (Logger Pro) that displays the data. I found the most economical and friendliest way for students to see the data from probe-ware is to use Vernier's LabQuest Mini interface hooked up to a computer with Logger Pro. LabQuest Mini has multiple ports, which is needed in many labs. They are the least expensive, so they are better on the budget. They require no batteries, so they are easy to transport if you need or want to. The other interfaces are more expensive, require batteries if you are going outside, and the stand-alone devices have a smaller screen to see the data, with less flexibility to manipulate the parameters like changing the time of data collection or changing units to modify an experiment. There are wireless probes and interfaces (that cost more) that may be easier to use if you do not mind the cost. A computer screen is much bigger to see the data physically, so this is my preferred setup. But using any interface will work fine for these labs.

Connecting the Probe-ware

To hook them up, you will plug your probe into one of the channels or the sonic on the interface. If the plug does not fit in smoothly, either you are plugging it in upside-down or at the wrong port. Then take the little chord that looks like it would go into your phone and plug that into your interface. Take the other end, and plug it into a USB port in your computer. Open up Logger Pro on your computer. If everything is hooked up properly and the computer and interface are working properly, you will see a green button at the top of the computer screen that says "Collect." Many of the labs have preset settings in Logger Pro. You will use the manila folder at the top left of the toolbar in Logger Pro to find the folders and files you will be instructed to go to for these specific settings for different labs. Whenever you get the physical equipment, they will have detailed instructions in the box on how to hook them up if you are still confused. They will also have instructions on how to calibrate the probes if needed. A few probes require frequent calibration; if we use any, it will be discussed in the lab directions. The more you use probe-ware, the easier it gets to set up. I usually only have to show my students twice for them to be able to set the equipment up on their own. But as you are showing them, have them physically do it. You can also find detailed instructions online at Vernier.com. Many more detailed labs can also be found there under lab ideas.

You also can use standard equipment like spring scales for force sensors or thermometers for temperature probes. Because schools want to integrate more technology, We wrote these labs to use probe-ware wherever applicable. Because they are so simple, these labs can be modified to fit whatever equipment you have. There are very few labs that I have used in my career that I did not modify how I presented them. One reason we wrote these labs this way was to customize them to the Texas TEKS and National Standards. We also wrote them the way we thought teachers would want to use them.

The Story of Life

This manual's sequence follows the Story of Life told by science. The content was placed to follow a storyline of how life evolved. The biochemistry and characteristics of life are mentioned first, then background information for ecology and evolution follows next. Each unit after goes over content as life evolved, first bacteria appeared, then single-celled eukaryotes, followed by multicellular organisms and sexual reproduction, plants, and multicellular heterotrophs and their body systems. Using life's evolution as the underlying theme helps students follow the process science shows brought life to this Earth and when much of the biology content would have appeared.

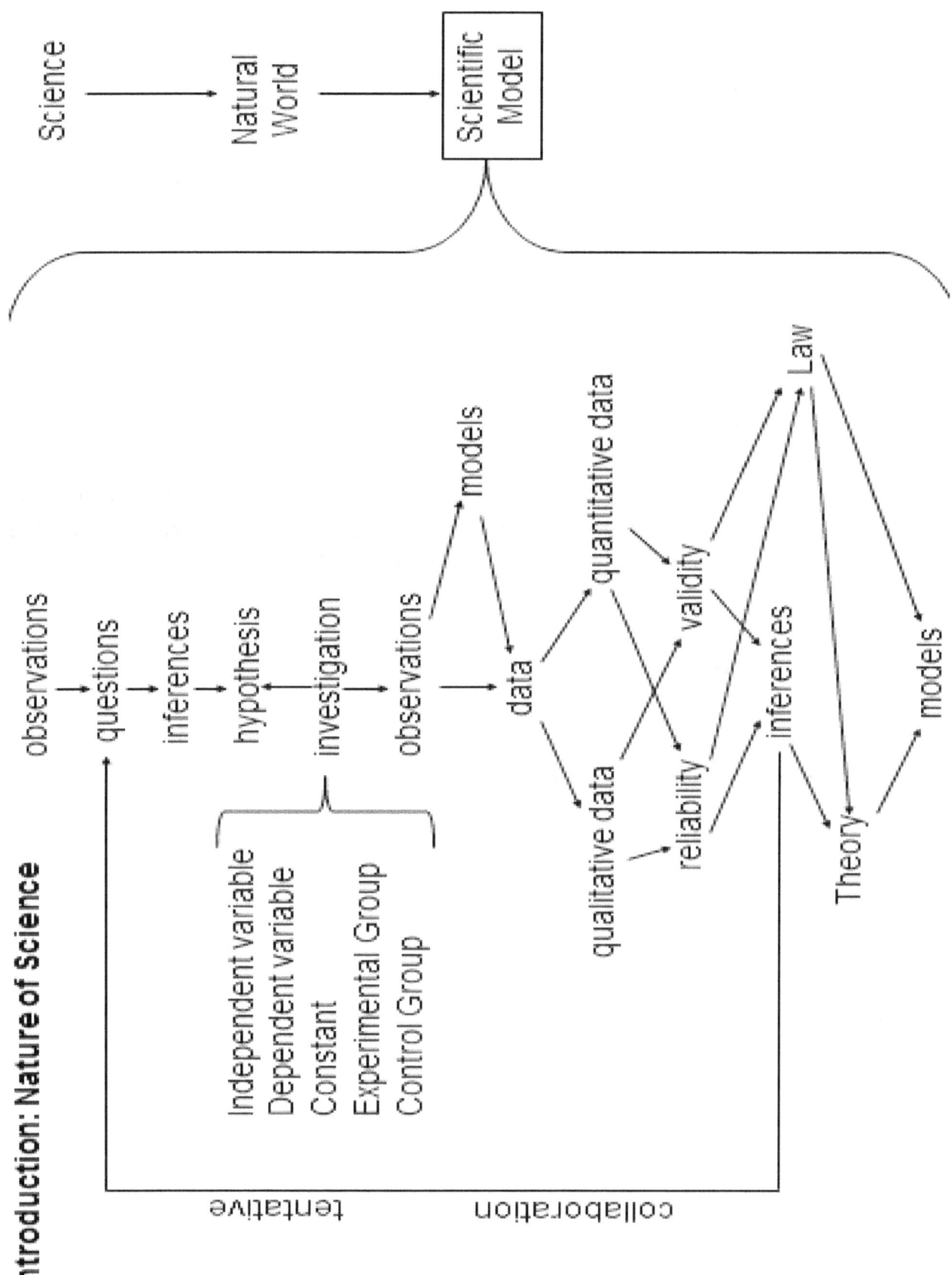

Introduction: Nature of Science

Focus on the Process

Directions:

Get a **small Legos set**. Make sure it is not too easy for your students. You are going to try to put it together in two different ways. Time how long it takes to put it together each way and answer the questions that follow. **Looking at the materials and lab we will be using, what are the safety precautions we should take to protect ourselves and materials during the investigation?**

A) Take the Lego pieces and try to construct the picture (the **product**) on the box's cover. Look at nothing but the front cover and the Lego pieces.

B) When 20 minutes have passed, or you are done, take what you have made totally apart. Take out the directions (the **process**) and construct the product while using the step-by-step directions. Make sure to time how long it takes to build the set.

Questions:

1) How did it feel to try to construct the Legos (A) without any directions?

2) Did you finish? If so, how long did it take?

3) How did it feel to construct the Legos (B) with the step-by-step directions?

4) Did you finish? If so, how long did it take?

5) Which strategy (A or B) allowed you to complete the product?

6) Which strategy (A or B) was more intimidating?

7) Which strategy (A or B) allowed you to see what is under the surface?

8) Which strategy (A or B) will allow you to learn more?

We often get anxious or procrastinate when faced with a large task. We are tempted to take a "shortcut" (copy or cheat, we do not learn much when we do this). There are pain and stress hormones that are released when this happens. One way to overcome this is to just worry about the next step in the process and not worry about the product. You can see and measure progress, which makes the process not feel too bad. Another way is to just start working. When you start working, those pain and stress hormones stop getting released so that anxiety goes away; this is why when we want to learn efficiently and effectively, we must:

Focus on the _____ and the _____ will take care of itself.

9) How is putting the Lego pieces together like putting ideas together to understand concepts?

Measurement Lab

Directions:

You will need **water**, a **scale**, a **meter stick**, a **temperature probe** attached to an **interface** connected to a **computer** with **Logger Pro**, a **100 mL graduated cylinder**, and a **stopwatch**. **Looking at the materials and lab we will be using, what are the safety precautions we should take to protect ourselves and materials during the investigation?**

1) Take the graduated cylinder and find its empty mass; write this in Data Table 1.
2) Add 50 mL of water to the graduated cylinder; make sure you use the meniscus properly where your volume is at the bottom of the meniscus. Have the teacher check that you measured it properly. Have each person in your group empty and fill the graduated cylinder with 50 mL of water. As they do so, have each person time how long it takes for each person to fill the graduated cylinder and check it is correct (it is not a race, just a chance to get familiar with using the graduated cylinder and stopwatch).
3) Now find the mass of the graduated cylinder with 50 mL of water in it. Subtract the empty graduated cylinder's mass from the full and write the water's mass in Data Table 1 below.
4) Hook your temperature probe up to an interface and hook your interface up to a computer with Logger Pro (unless you have a LabQuest 2, then just hook your probe to the LabQuest 2). Find where the Logger Pro is located on your computer so you can use it again in the future. Once open, find the graduated cylinder's water temperature in both Fahrenheit and Celsius (you will have to figure out how to change the units). Write these in Data Table 1.
5) Take your meter stick and measure the length of the graduated cylinder. And measure the width of the base in centimeters. Write these in Data Table 1.

Data Table

Object	Mass (g)	Volume (mL)	Time to Fill (s)	Temp (°F)	Temp (°C)	Length (cm)	Width (cm)
Graduated Cylinder		✖		✖	✖		
Water			✖			✖	✖

Questions:

1) Convert a length to meters, the volume to liters and a mass to kilograms, and Celsius to Kelvin.

   Length _____ m   Volume _____ L   Mass _____ kg   Temp _____ K

2) What do you notice about the mass of the water compared to its volume?

3) What can happen to your investigations if your measurements are not accurate or precise?

4) Why do you think the rest of the world uses the metric system over the English system?

Patterns in Pennies

Directions:

You will need a **ruler**, 10 **pennies**, a **scale**, a **roll of pennies**, and an **empty penny roll. Looking at the materials and lab we will be using, what are the safety precautions we should take to protect ourselves and materials during the investigation?**

1) Find the mass of one penny with a scale to the nearest .1 g. Then measure the height of the penny in millimeters. Write these in Data Table 1 below.
2) Place another penny on top of the original penny and find the mass and height of the two pennies. Write these in Data Table 1 below.
3) Keep adding pennies one by one, measuring the mass and height until you have 10 pennies on the scale.
4) Make a line graph with the mass on the (*x*) axis and the height on the (*y*) axis for the pennies on Graph 1.

Data Table 1

Number of Pennies	Mass	Height
1		
2		
3		
4		
5		
6		
7		
8		
9		
10		

Graph 1

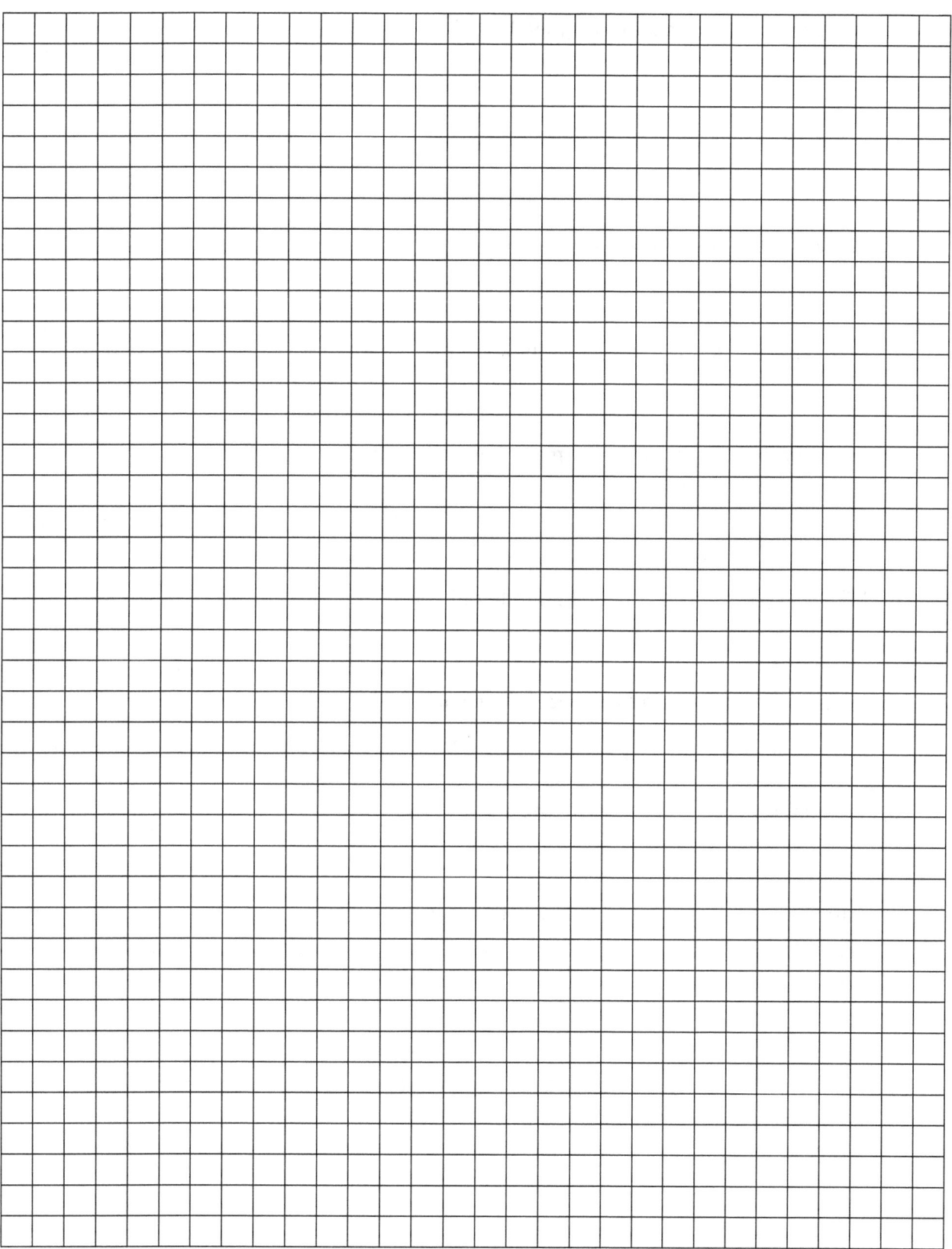

Questions:

1) What do you notice about the graph?

2) Is this a direct or inverse relationship between mass and height?

3) Do all pennies have the same mass? (Explain)

4) Do all the pennies have the same thickness? (Explain)

5) Use your data to estimate how many pennies are in the coin roll. How many pennies do you think are in the roll?

6) What did you do to estimate the number of coins?

7) What else could you do to estimate the coins?

8) Try that. Do you get the same number as #5?

9) Carefully open up the coin roll and find out how many pennies there are. How close were you to the real number? After being done counting, carefully close the roll back up.

10) Calculate the % accuracy by taking the lowest number between your guess and the actual number dividing by the higher of the two, then multiplying by 100.

11) What were some sources of error?

Virtual Investigations that go with Introduction

ExploreLearning.com:

Unit conversions Gizmo

Solving Using Trend Lines Gizmo

Graphing Skills Gizmo

Elevator Operator Gizmo

Measuring Volume Gizmo

Weight and Mass Gizmo

Triple Beam Balance Gizmo

Reaction Time 1 Gizmo

Reaction Time 2 Gizmo

Physicsclassroom.com:

Concept Builders:

Relationships and graphs

Experiments and Variables

Proportional Reasoning

Using Graphs

Which One Does Not Belong

Chemistry

Significant Digits and Measurement

Metric System

Metric Estimation

Unit 1 Nature of Life

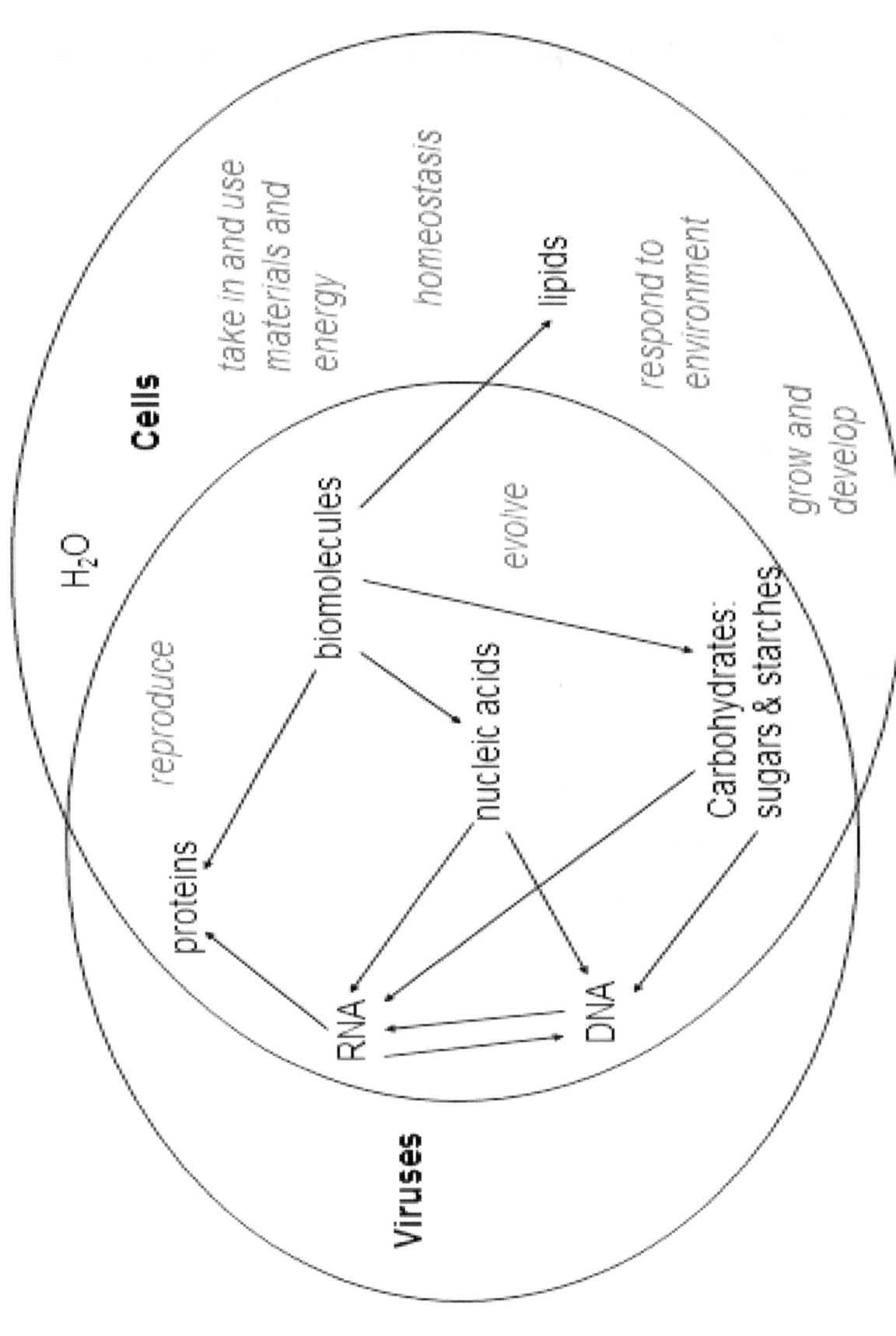

Scale Model of a Hydrogen Atom

Directions and Questions:

You will need a **golf ball**, a **bead**, and a **large field** or **parking lot. Looking at the materials and lab we will be using, what are the safety precautions we should take to protect ourselves and materials during the investigation?**

1) Walk out to a large field or parking lot, at least the size of a football field. Keep in mind the space you use still may be too small to be a scale model. You will make a model of a hydrogen atom with 1 proton and 1 electron.

2) On one edge, take a small red bead representing an electron and put that somewhere where you can see it (hang it on a fence or a tiny branch).

3) Walk at least 100 yards away; if you have more room, you can use that. Hold up the golf ball, which is a proton, and answer the questions that follow. Can you see the bead?

4) This distance is how far away the closest electron speeds around the proton if they are this size. The speed approaches the speed of light. It moves so fast that it makes a ball the size of a football stadium. If you have ever seen a fan moving fast, does it look like a disk? But is it a disk?

   a. So we have to ask ourselves, the atom looks like a ball, but is it a ball? Explain why.

5) This illusion is the same type of illusion. When an object moves close to the speed of light, time stops for that object. Quantum physics allows objects without time to be anywhere they would normally be at any time at the same time; this is why the electron is said to be everywhere in the electron cloud at the same time. What do you see between the proton and electron?

6) What would happen to the electron if time were to stop?

7) If that would happen to the electron, what would happen to the atom?

8) What would happen to all atoms in the area where time stops?

9) This relationship is why we call this the space-time continuum. You cannot have space without time; this is how black holes form. When time disappears, so does space. What color is a black hole? Explain why.

10) Is there any space at the bottom of the black hole?

11) This concept also shows that the pixel of our universe is an illusion. If the pixel of our universe is an illusion, what does that say about our universe?

12) What do you think is the ultimate reality?

13) With our laws of physics, are we allowed to know?

Building Bohr Models

Directions:

You will need a **film case** filled with **three colors of beads** (I use red, white, and blue) and a **periodic table. Looking at the materials and lab we will be using, what are the safety precautions we should take to protect ourselves and materials during the investigation?**

1) Designate which color bead represents which part of the atom. Here is an example: Red beads are electrons, white beads are neutrons, and blue beads are protons. I use these colors because this is how my periodic table was colored, and these colors help my students learn how and why the periodic table works.

2) Carefully empty the film case. As the students empty those on the table, some beads will bounce out. I call this radiation because radiation is particles and energy coming out of an atom.

3) Also know that a neutron is the combination of a proton and an electron. This is why it is neutral. Proton (p+) + Electron (e-) = Neutron (n°) or 1+ + 1- = 0.

4) Place the empty film canister in the center circle labeled the nucleus.

5) Now make a **Hydrogen atom** by looking at the periodic table and having the student look at the atomic number; this tells us how many protons are in that atom. The number of protons is what defines the element. Hydrogen's atomic number is one, so place one blue proton in the nucleus (film case). This number also tells you how many electrons there will be in a neutral atom. So place one red bead in the first orbit closest to the nucleus. Protons and neutrons have mass, and electrons do not. Look at the average atomic mass at the bottom of the element's box. Round it to the closest whole number; this tells you the mass of most of the atoms of that element. Since the mass is one and we have one bead in the nucleus, we have completed the hydrogen 1 (H-1) isotope.

6) Add a neutron (white bead) to the nucleus; this makes a Hydrogen 2 (H-2) isotope. One proton and one neutron in the nucleus (2 beads) give us a mass of 2. Hydrogen 2 isotope is another version of the same element.

7) Next, make a **Helium atom**. Look at the periodic table and find Helium. How many protons does it have? Add a blue bead to the nucleus to give it 2 protons.

8) How many electrons does Helium need to have? Add one red bead to the first circle but on the opposite side from the other electron.

9) Why would the electron be there?

10) Now, look at the mass of the Helium. How many does it have (remember to round to the whole number)?

11) How many beads are in your nucleus?

12) Add white beads until you have the right amount of beads in the nucleus (film canister) as the average mass; this makes a Helium 4 (He-4) isotope.

13) Now take one neutron (white bead) out to make Helium 3 (He-3) isotope; this is what is on the moon, and scientists will want this in the future to make fusion reactions to produce electricity.

14) We have now filled the first energy level of electrons. Notice we are on the right end of the periodic table. No more electrons can go into this circle.

15) We will now make a **Lithium atom**. How many protons does it have? Place a blue bead in the nucleus (film canister) until you match that number of protons.

16) How many electrons does Lithium have? Place an electron (red bead) on the second circle. We are now at the second energy level.

17) Look at the average mass of Lithium. Add neutrons (white beads) to the nucleus until you have matched Lithium's mass with the number of beads in the nucleus (film canister).

18) Now go and make the other isotopes, **Carbon 12** (which is in all life) and **Carbon 14** (which is radioactive and starts to decay when an organism dies), which are important to life. Carbon 14 turns into **Nitrogen** by taking an electron out of a neutron (a proton and an electron); this leaves a proton in the nucleus, making it a Nitrogen atom.

19) Then build **Oxygen**, **Sodium**, **Magnesium**, **Phosphorous**, **Potassium**, and **Calcium**. These are all elements that are important in large numbers inside of life.

Bohr Model

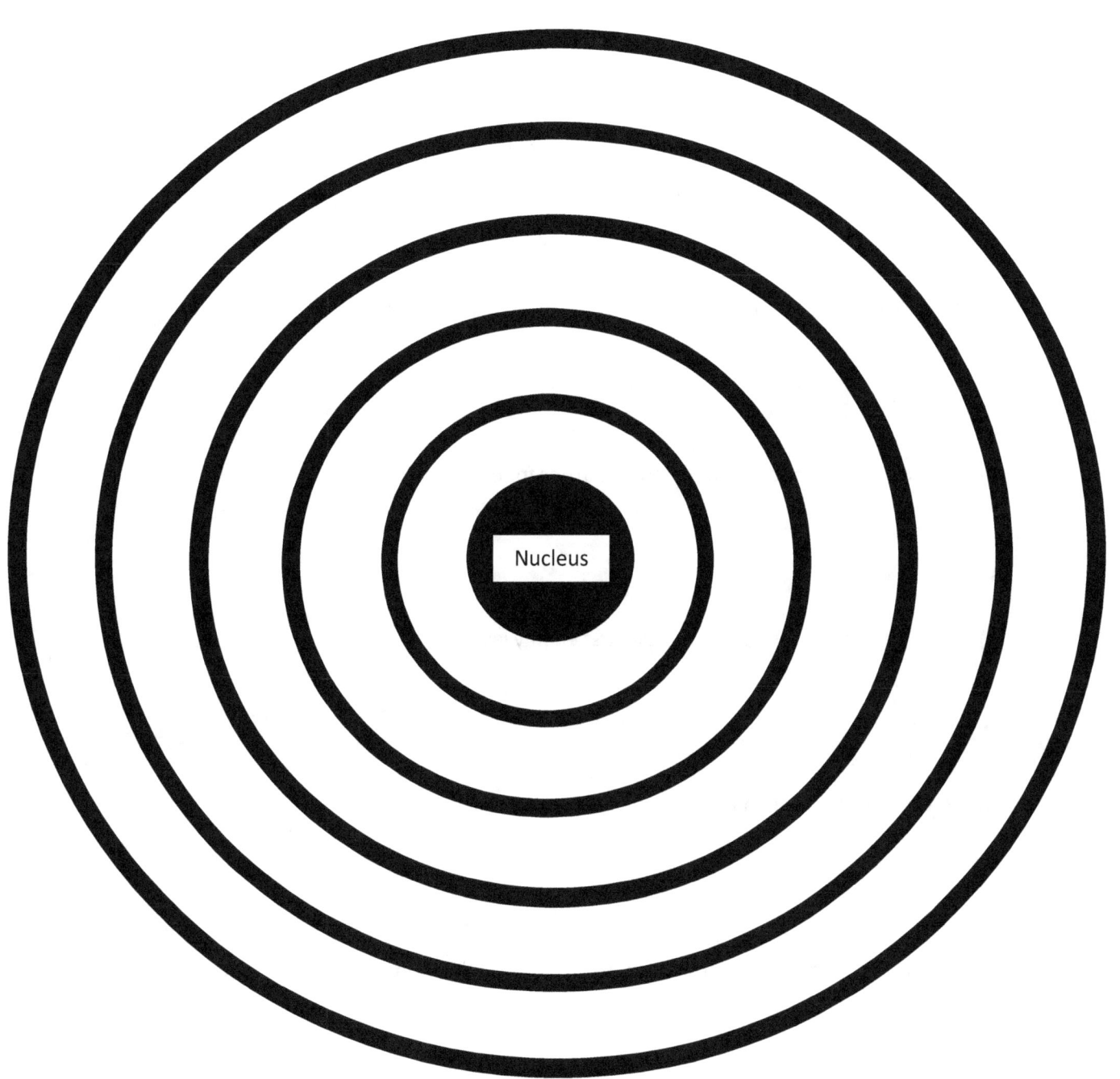

Questions:

1) How is the atomic mass determined?

2) What part of the atom determines what element it is?

3) How do you find out the number of neutrons for an element?

4) How do you find out the number of electrons for each element?

5) How many electrons can go into the first energy level?

6) How many electrons can go into the second energy level?

7) How is the periodic table structured to tell us about the atoms of each element?

Models of Micro-molecules

Directions:

You will need a **molecular model kit** and a **Periodic Table. Looking at the materials and lab we will be using, what are the safety precautions we should take to protect ourselves and materials during the investigation?**

1) At the top of your periodic table, label it like this just below:

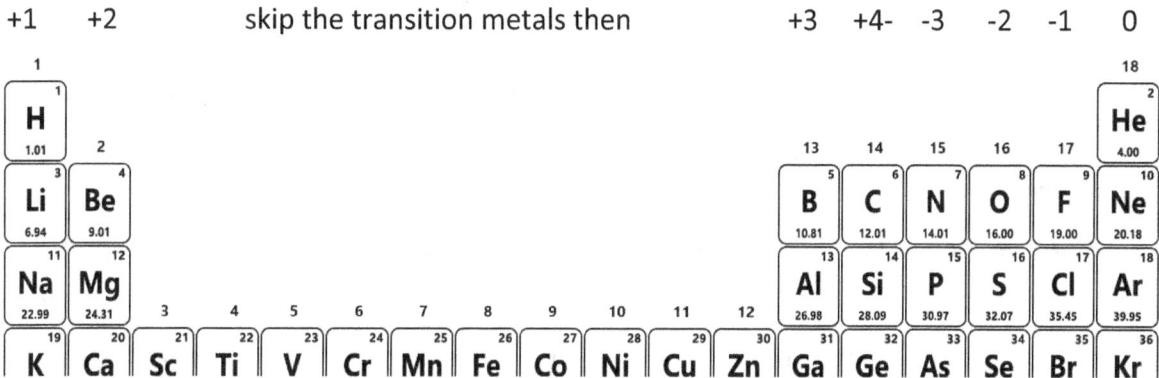

2) Different kits have different colors. In my kit, the:

    a. +1(one-prong white) represents the Alkali Metals

    b. +2 (two-prong yellow) represents the Alkaline Earth Metals

    c. +3 (three-prong blue) represents the Boron Group

    d. +/- 4 (four prong black) represents the Carbon Group

    e. -3 (three-prong red) represents the Nitrogen Group

    f. -2 (two-prong blue) represents the Oxygen Group

    g. -1 (one-prong green) represents the Halogens

    h. The white tube is the bond

3) The different pieces in #2 represent the elements in those groups. Put the following molecules together:

H_2 (Make two): Take two hydrogen (white one-pronged) pieces connected by one white tube (single bond); this is the simplest molecule made and the most abundant in the universe.

O$_2$ (Make two) Take two oxygen (small blue two-pronged) pieces connected by two white tubes (double bond); this is **oxygen gas**. It makes up 18-20% of our atmosphere and is needed for aerobic respiration.

H$_2$O Take the two hydrogen molecules and the two oxygen molecules and rearrange them to make two **water molecules**. The oxygen (blue two-prongs) in the middle is connected with two white tubes to the two Hydrogen (white one-pronged) pieces (two single bonds); this is how rocket fuel burns to make water. If you do this in the opposite direction, this is how water separates during photosynthesis or electrolysis when electricity hits the water, and hydrogen and oxygen separate into their gases.

N$_2$ Take two Nitrogen (red three-pronged) pieces and connect them to each other with three tubes (triple bond). This molecule is nitrogen gas, the most abundant element in our atmosphere (78-80%).

NH$_3$ Take the one Nitrogen (red three-pronged) piece and connect it to three hydrogens (one-pronged white) pieces by three tubes (three single bonds); this is **ammonia**.

CO$_2$ Take one carbon (black four-pronged) piece and connect it to two oxygen (blue two-pronged) pieces with four white tubes (two double bonds); this is **carbon dioxide**.

CH$_4$ Take one carbon (black four-pronged) piece and connect it to four hydrogens (white one-prong pieces) with four white tubes (four single bonds); this is **methane gas**, the base for all fuels. **Methane** is a byproduct of decomposition away from oxygen. It is in all farts. Methane is also found in abundance away from Earth in the universe.

Questions:

1) What determines how atoms will combine with other atoms?

2) When building the molecules, what do you notice about each of these molecules?

3) How many atoms are in the largest molecule we made? This size is why we call them micro-molecules; they are very small.

Models of Macromolecules

Directions:

You will need a **molecular model kit**, a **Periodic Table**, and a **textbook** or **internet**. **Looking at the materials and lab we will be using, what are the safety precautions we should take to protect ourselves and materials during the investigation?**

1) At the top of your periodic table, label it like this just below:

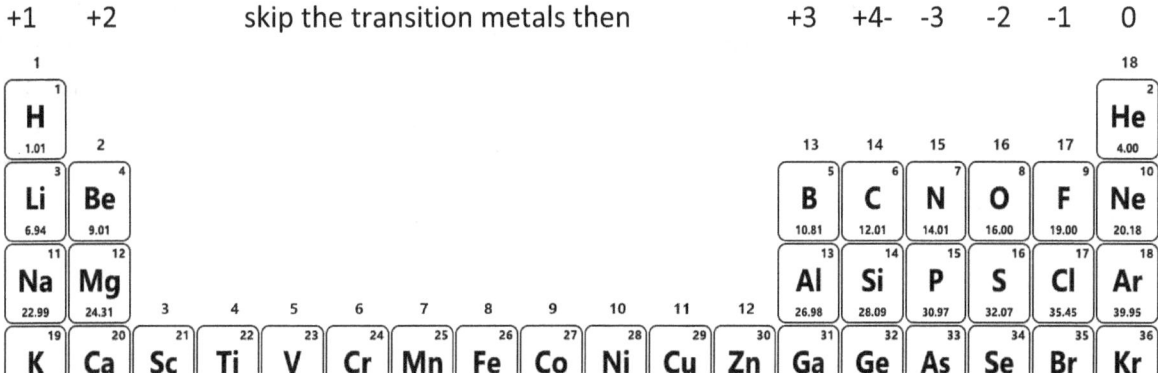

2) Different kits have different colors. In my kit, the:

    a. +1 (one-prong white) represents the Alkali Metals
    b. +2 (two-prong yellow) represents the Alkaline Earth Metals
    c. +3 (three-prong blue) represents the Boron Group
    d. +/- 4 (four prong black) represents the Carbon Group
    e. -3 (three-prong red) represents the Nitrogen Group
    f. -2 (two-prong blue) represents the Oxygen Group
    g. -1 (one-prong green) represents the Halogens
    h. The white tube is the bond

3) The different pieces in #2 represent the elements in those groups. Use pictures of large molecule monomers from your textbook or the internet to help you put them together. Then follow the directions to put monomers together to make polymers.

    a. **Carbohydrates**: the **monomer** is sugar like **glucose**. A **polymer** could be **starch**. When you make a chain of sugars by taking off a hydrogen atom from one sugar and OH from the other, this will produce water. You can then put the two sugars together to start a chain. This process is called **dehydration synthesis**. When you separate the starch back into sugars (like in digestion), water gets taken apart, and the H and OH get put back into each sugar molecule; this is called **hydrolysis**. Model this with your group. Sugar is used as fuel in living things.

b. **Lipids**: **Glycerol** and **fatty acids** are the **monomers**. They make membranes and store energy.

c. **Protein**: **Amino acids** are the **monomers** put together with **peptide bonds,** the building blocks of cell parts, enzymes, and hormones. An average usable protein is 300-400 amino acids long.

d. **DNA & RNA**: The **monomers** are **nucleotides** made of **sugar, phosphate**, and a **nitrogenous base**. You have 6 feet of DNA in one cell. DNA is the genetic material for storing life's information to help build protein chains. Three types of RNA: **mRNA** (messenger RNA), **tRNA** (transfer RNA), and **rRNA** (ribosomes), all of which are needed to build proteins. If you are missing any of these, protein cannot be made, so; life cannot be made.

Questions:

1) From what you have observed in this lab, what are monomers?

2) What are polymers?

3) How many atoms are in the smallest molecule you made in this lab?

4) Can you count how big the largest would be?

5) If we were to build a cell with these model pieces, how big do you think it would be?

6) Is the chemical organization of life simple or complex? Explain.

7) Fill in the chart below to compare and contrast the macromolecules.

Polymers	Carbohydrates	Lipids	Protein	DNA & RNA
Monomers				
Functions				
Pictures				
What does it contain? Elements Calories/gram				

8) How do these models give evidence to the abiogenesis hypothesis?

     a. Why is abiogenesis a hypothesis?

9) Compare and contrast the abiogenesis hypothesis with the Law of Biogenesis.

     a. Why is this biogenesis a law?

10) Why aren't the abiogenesis hypothesis and the Law of Biogenesis theories?

DNA Extraction

Directions:

You will need a **100 mL graduated cylinder**, a **scale**, a **spatula**, a **Ziplock bag**, a **small beaker**, three **strawberries**, 5 mL **dishwashing liquid**, 50 mL of **water**, 5 g of **salt**, a small **wire strainer**, 10 mL of **alcohol**, and a **toothpick** or **Popsicle stick**. **Looking at the materials and lab we will be using, what are the safety precautions we should take to protect ourselves and materials during the investigation?**

1) Measure 50 mL of water and pour it into the small beaker. Measure 5 mL of dishwashing liquid and pour it into the same beaker.

2) Measure 5 g of salt with your scale and use the spatula to move the salt into the same small beaker.

3) Take your three strawberries and place and place them into the Ziplock bag. Pour the beaker's contents into the Ziplock bag on top of the strawberries. Seal the bag shut and carefully squish the strawberries into a paste without breaking the bag.

4) Rinse out your beaker, put the small wire strainer on top of the beaker, and pour the bag's mashed contents onto the strainer to let the liquid contents go through into the beaker filling it ½ to ¾ full. Your DNA sample is now dissolved in the liquid in the beaker.

5) Slowly add 10 mL of alcohol into the beaker to bring it out of the solution.

6) Take your toothpick or Popsicle stick and gently move it around in the beaker to gather the fibers of DNA you see.

Questions:

1) Where did the DNA originate?

2) What released the DNA from its cells (the cell membrane is made of lipids)?

3) Do you think this investigation would work with any other foods? Explain why.

4) How could we adapt this investigation to extract DNA from leaves or mushrooms?

5) Could this method work for animal tissue like meat or liver? Explain.

6) If you have a **blender** and different foods, try this investigation again (have each group try a different food). Why do you think we should find DNA in other fruits, vegetables, meat, and mushrooms?

7) Where do you think we get the materials to replicate our own DNA for our bodies?

Practice Reading Nutrition Labels

Directions and Questions:

You will need the **internet** or a **textbook**, and **nutrition labels** from various foods, at least 3 per student.

1) Which biomolecules do you see in each food?

2) Which food had the most fat?

    a. How does the body use fat/lipids?

3) Which food had the most Carbohydrates?

    a. How does the body use carbohydrates?

4) Which food had the most protein?

    a. How does your body use protein?

5) Which food had the most sugar?

6) Which food had the most sodium?

7) Which food had the most cholesterol?

8) Which food had the most vitamins and minerals?

9) Which food here seems to be the most nutritious? Explain why.

10) Which food here seems to be the least nutritious? Explain why.

Model of Denaturing an Enzyme

Directions:

You will need 30 **pennies**, a **stopwatch**, a **tennis ball**, and some type of **tape. Looking at the materials and lab we will be using, what are the safety precautions we should take to protect ourselves and materials during the investigation?**

1) Separate all 30 pennies so they are lying flat on the floor.
2) Have one person act as the enzyme, one person keep track of the time, and one person collect the data for others to copy later.
3) The enzyme will have 10 seconds to pick up as many pennies as possible with one hand, one at a time, placing those heads up on the table/desk. Count how many pennies you placed heads up and write this data in Data Table 1. Place the pennies back on the floor.
4) Repeat the procedure in #3 four more times.
5) Now tape the thumb of the person acting as the enzyme to the palm of their hand. Repeat the procedure in # 3. Taping the thumb is modeling a form of a denatured enzyme.
6) Take the tape off of the thumb holding it to the palm. Now tape a tennis ball to the palm of their hand to test the effects of placing an inhibitor in the activation site of the enzyme. Repeat the procedure in # 3.

Data Table 1

Trials (10 seconds each)	Normal Enzyme picked up how many pennies	Denatured Enzyme (taped thumb) picked up how many pennies	Inhibited Enzyme (tennis ball) picked up how many pennies
1			
2			
3			
4			
5			
Total			

7) Graph the data in Data Table 1 on Graph 1 below.

Graph 1

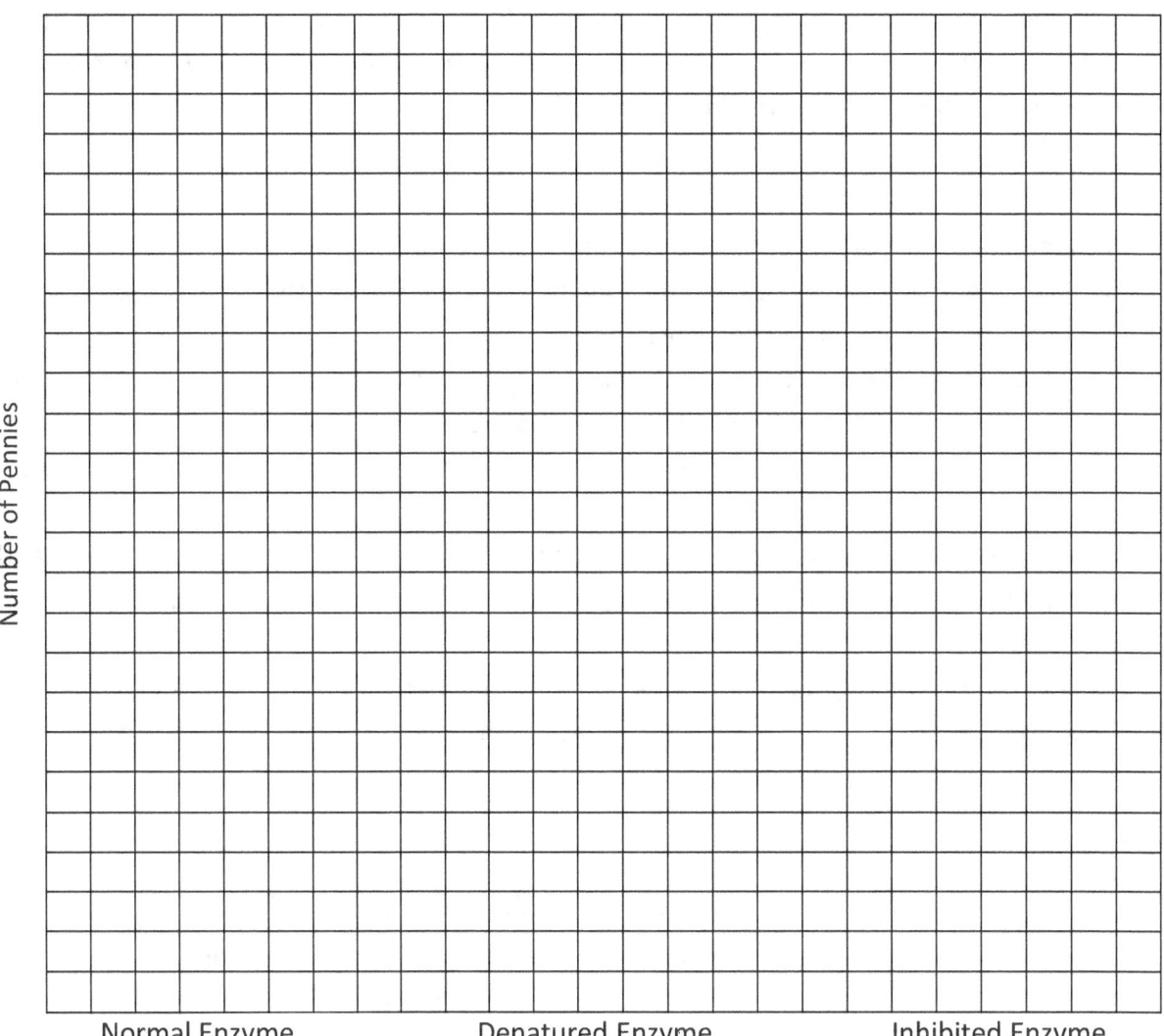

Number of Pennies

Normal Enzyme   Denatured Enzyme   Inhibited Enzyme

Questions:

1) Which object represented the enzyme?

2) Where was the active site?

3) How was the activation site denatured?

4) What was the inhibitor?

5) What was the substrate?

6) How did denaturing the enzyme affect how many pennies were processed?

7) How did inhibiting an enzyme affect how many pennies were processed?

8) How can enzymes be denatured?

9) What do you think are some real inhibitors?

10) What could happen to a person if the enzymes that make insulin (an enzyme that helps regulate blood sugar) for your body are inhibited or denatured?

Virtual Investigations that go with Characteristics of Life

ExploreLearning.com

Element Builder Gizmo

Identifying Nutrients Gizmo

Building DNA Gizmo

Chemical Equations Gizmo

Enzymes STEM Case Gizmo

Enzymes Handbook Gizmo

PhET.colorado.edu

Molecular Polarity

Reactions and Rates

Salts and Solubility

Sugar and Salt Solutions

Notes:

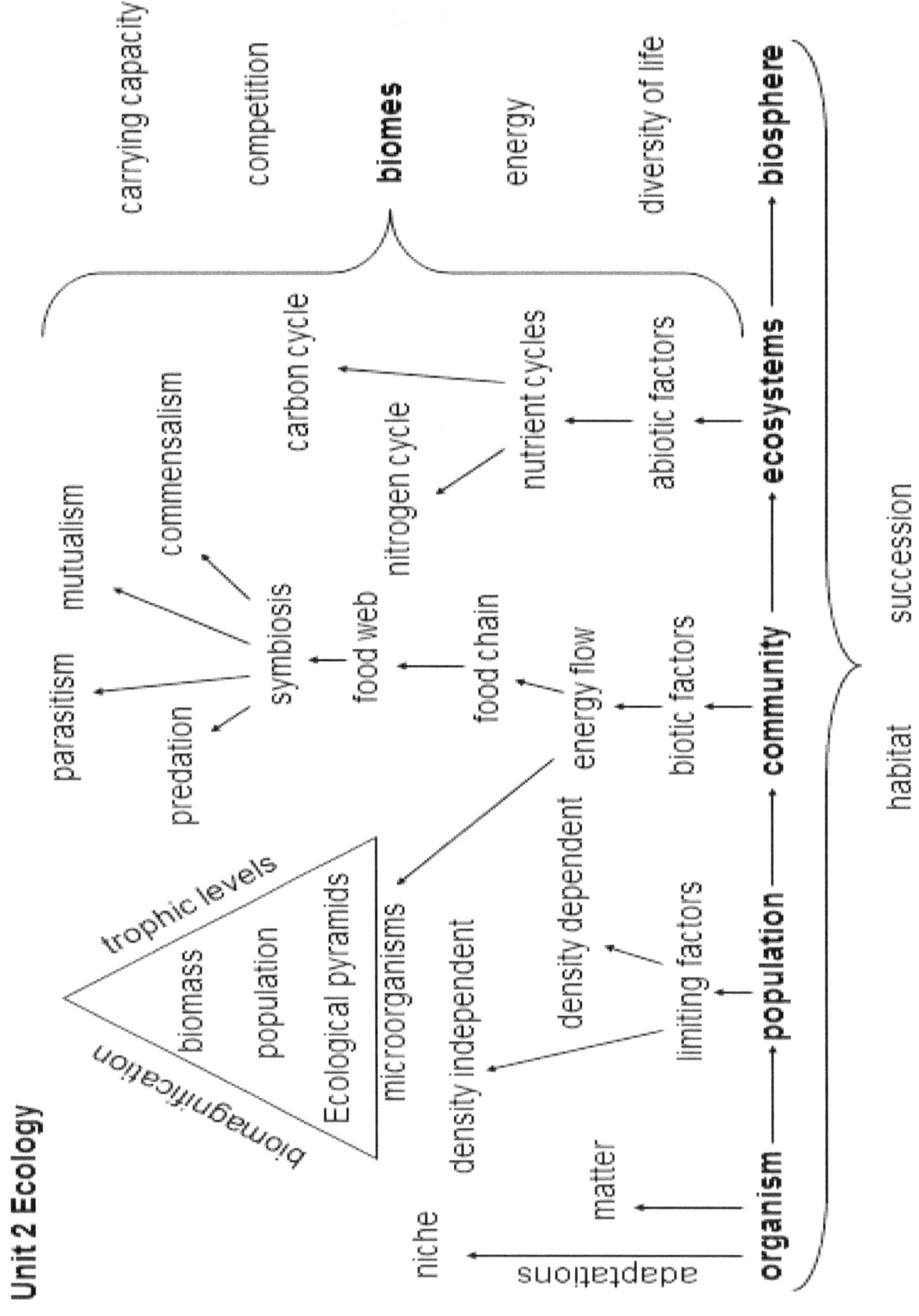

Unit 2 Ecology

Environmental Issues and Ethics

Directions: Use the **internet** and **textbook** to help you fill in the chart about environmental policies.

Legislation, Treaty, or Protocol	What is it about?	Who enacted it?	Where is it in effect?	Why was it made?
Clean Air Act				
Clean Water Act				
Soil and Water Resources Conservation Act				
Texas Automobile Emissions Regulations				
National Park Service Act				

Endangered Species Act			
Environmental Arctic Treaty System			
Montreal Protocol			
Kyoto Protocol			
Paris Accord			

1) What do you think is the purpose of laws, treaties, and protocols?

2) What careers would be interested in these laws, treaties, and protocols?

Symbiotic Relationships

Directions and Questions:

1) Read how the different types of symbiosis (the interaction relationship between organisms) can be defined.
   a. **Mutualism** can be defined as two organisms interacting where both benefit. **+ +**
   b. **Commensalism** can be defined as two organisms interacting where one benefits and the other is neither harmed nor benefits. **+ o**
   c. **Parasitism** can be defined as two organisms interacting where one benefits and the other is harmed. **+ –**
   d. **Predation** can be defined as where one benefits by eating the other. **+ –**
2) Use these definitions to classify relationships between organisms interacting below and tell why they have that type of relationship.
   a. A liver fluke causes disease in an animal.

   b. Lichens are a combination of organisms where the fungus absorbs minerals and give them to algae, and the algae go through photosynthesis to make sugar for energy and give it to the fungus.

   c. A Remora eats the scraps left behind from the feeding of a shark. The shark receives nothing in return.

   d. A dog gets fed, has a home, socialization with a family of humans. The humans get home security, lower blood pressure, the release of stress, entertainment, companionship, and a sense of importance when taking care of the dog.

   e. A robin chases and eats a cicada.

   f. A cow eats grass, and the grass gets digested by the bacteria living in the cow's digestive tract releasing the nutrients.

g. A bird eats the parasites in the mouth of an alligator.

h. Ticks and fleas live on a dog, eating its blood, thus causing the dog to itch.

i. A clownfish lives within the sea anemone. The fish gets protection and food, and the anemone gets rid of parasites as the fish eats them.

j. Mistletoe gets food and water from the tree, which gets damaged.

k. A bird makes a nest in the tree, and the tree does not get helped or harmed.

l. Tardigrades are microscopic animals that pierce plant cells and suck the contents out through their straw-like mouths.

3) Go outside and find an example of each you observe and write it below.
   a. Mutualism

   b. Commensalism

   c. Parasitism

   d. Predation

Competitive Relationships

Directions:

Use the picture below to identify the model ecosystem's biotic and abiotic factors. Then discuss with your teacher and class the questions that follow and write down your answers.

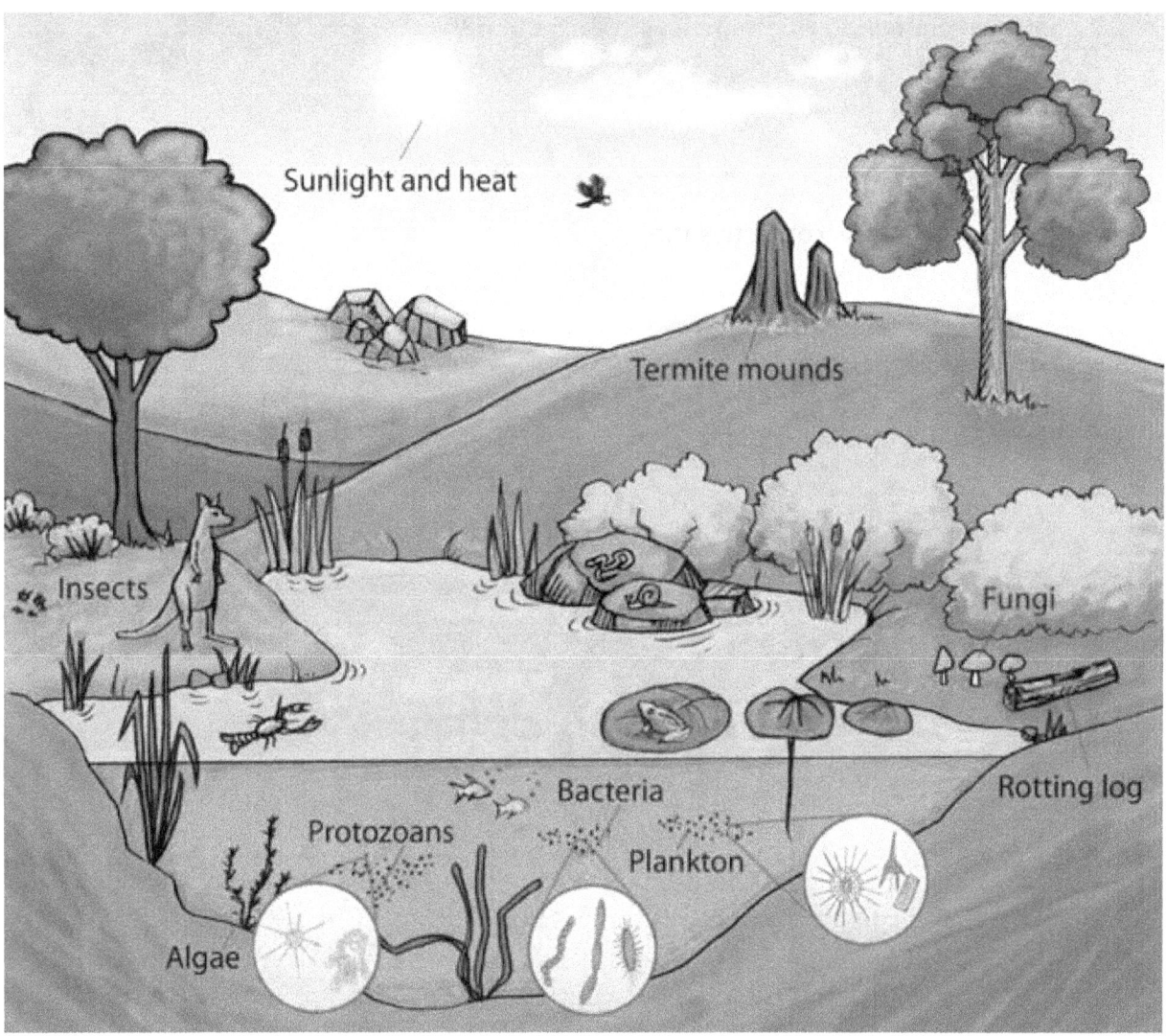

Picture from acamrmichael.weebly.com

Questions:

1) What are the biotic factors in this ecosystem?

2) What are the abiotic factors in this ecosystem?

3) Which organisms would compete with each other?

4) What would they be competing for?

5) Which organisms here would be competing for soil?

     a. How might they compete for this soil?

     b. How might plants avoid this competition?

6) Where might you find plants competing for light?

     a. How would they compete for the limited light available?

b. How might plants avoid this competition?

7) Where might we find plants and animals competing for water?

a. How would they compete for limited water opportunities?

b. How might plants avoid this competition?

8) How do animals compete for food?

a. How might animals avoid competition for a food source?

9) How could animals compete for a range of temperatures?

a. How might animals avoid competition for a temperature range?

10) How could plants compete for a range in temperature?

      a. How might plants avoid competition for a temperature range?

11) Why do you think no two species can occupy the same niche for a long period of time?

12) What happens when an invasive species enters a new ecosystem it was not in before?

Build an Ecosystem

Directions:

We are going to build a model ecosystem out of **Play-Doh**. Use the 10% rule to help you find the amount of Play-do you will need for each trophic level. Measure the amount of biomass with a **scale** for each level in your community (see the chart below). Then form the plants and animals that will be living in your community. Because the number will be so small, when you get to the quaternary consumer (red), it will be best for the teacher to make one animal for the whole class to migrate between all the small communities. After each group has built its community, the whole class will represent a larger community (the biotic parts of an ecosystem). **Looking at the materials and lab we will be using, what are the safety precautions we should take to protect ourselves and materials during the investigation?**

Data Table 1 (Group Data)

Play-Doh Color	Trophic Level	Amount Used (g)
Green	Producer (Plants)	200.00g
Blue	Primary Consumer (Herbivore)	g
Orange	Secondary Consumer (1st Level Carnivore)	g
Purple	Tertiary Consumer (2nd Level Carnivore)	g
Red	Quaternary Consumer (3rd Level Carnivore)	g
Black	Scavengers/Decomposers	5.00g

Data Table 2 (Whole Class Data)

Play-Doh Color	Trophic Level	Amount Used (g)
Green	Producer (Plants)	g
Blue	Primary Consumer (Herbivore)	g
Orange	Secondary Consumer (1st Level Carnivore)	g
Purple	Tertiary Consumer (2nd Level Carnivore)	g
Red	Quaternary Consumer (3rd Level Carnivore)	g
Black	Scavengers/Decomposers	g

Questions:

1) Why did we use the amounts we did for each tropic level?

2) Why are high-level predators very territorial?

3) Compare and contrast the pyramid of biomass, energy, and number of organisms.

4) What do you think determines how many decomposers an ecosystem will have?

5) Can an animal occupy more than one tropic level simultaneously in an ecosystem? Explain.

- Clean up by putting all the Play-Doh back in their original containers when we are done.

Ecological Pyramid

Instructions:

Side 1: Label food chain with the names of organisms.

Side 2: Draw pictures of the organisms.

Side 3: Label Trophic Levels: Producers, primary consumers, secondary consumers, ect.

Side 4: Starting with 14,800 Kcal, show how the energy flows with the 10% rule.

Put your name on it, cut out the model by cutting along the outer edge, fold the corners and flaps to make a pyramid. Add paste or glue to the flaps to hold it together.

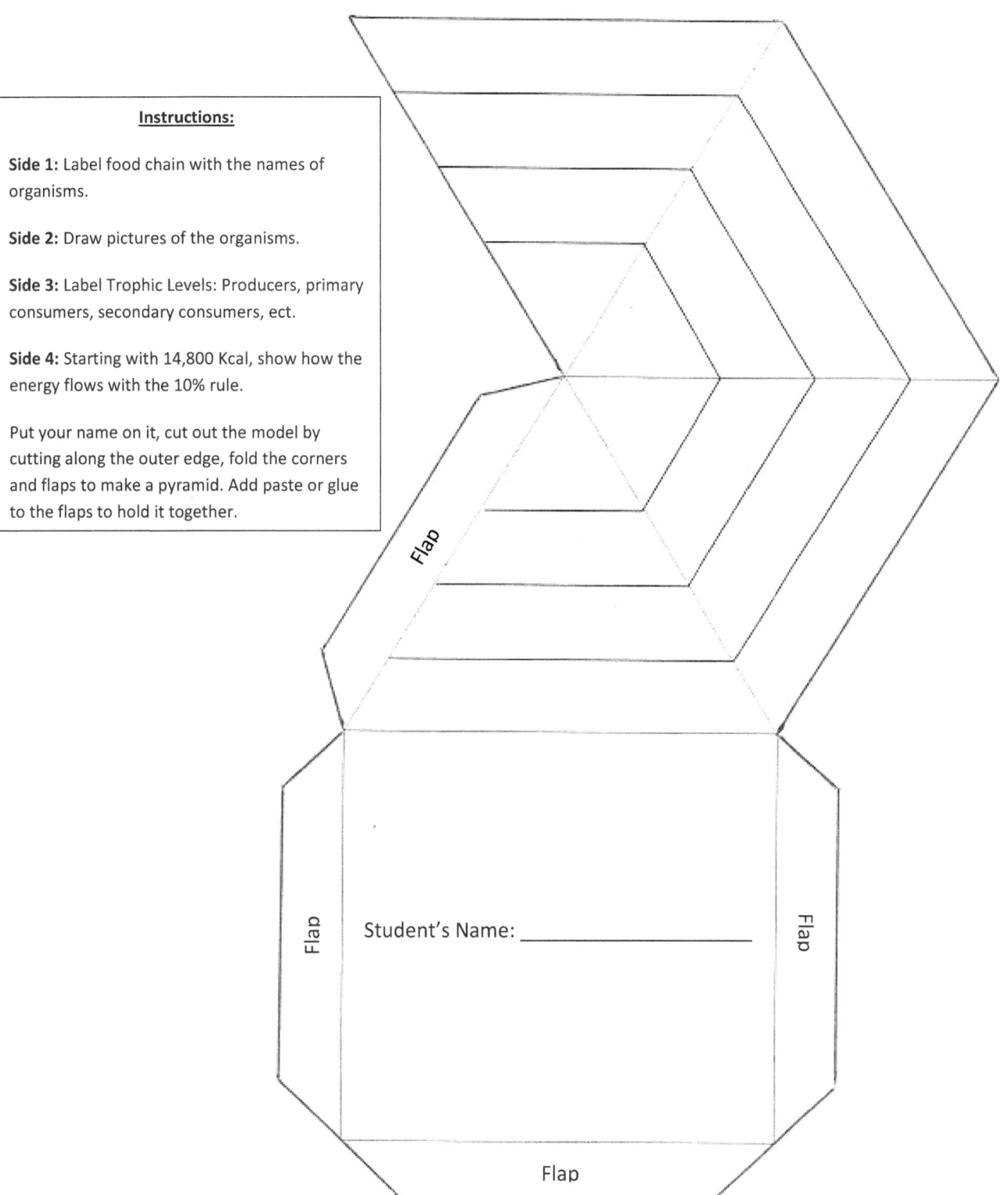

Flap

Flap

Flap

Student's Name: _____

Flap

This page will be cut from the previous page.

Social Behavior

Directions:

Use the **internet** and your **textbook** to research social behaviors and answer the following questions.

1) What is the difference between individual and social behavior?

2) Why do birds flock, fish school, and cattle herd? What benefits do these behaviors give them?

3) What competition happens in these behaviors?

4) What adaptations allow the speed at which birds react to a flock and fish react to a school?

5) Why is it beneficial for animals of the same group to cooperate while hunting? Give examples.

6) Why is it beneficial for animals of the same group to cooperate while migrating? Give examples.

What Happens to the Food Web?

Directions:

Use the food web below to predict what would happen to the populations of the organisms when a change happens to the ecosystem.

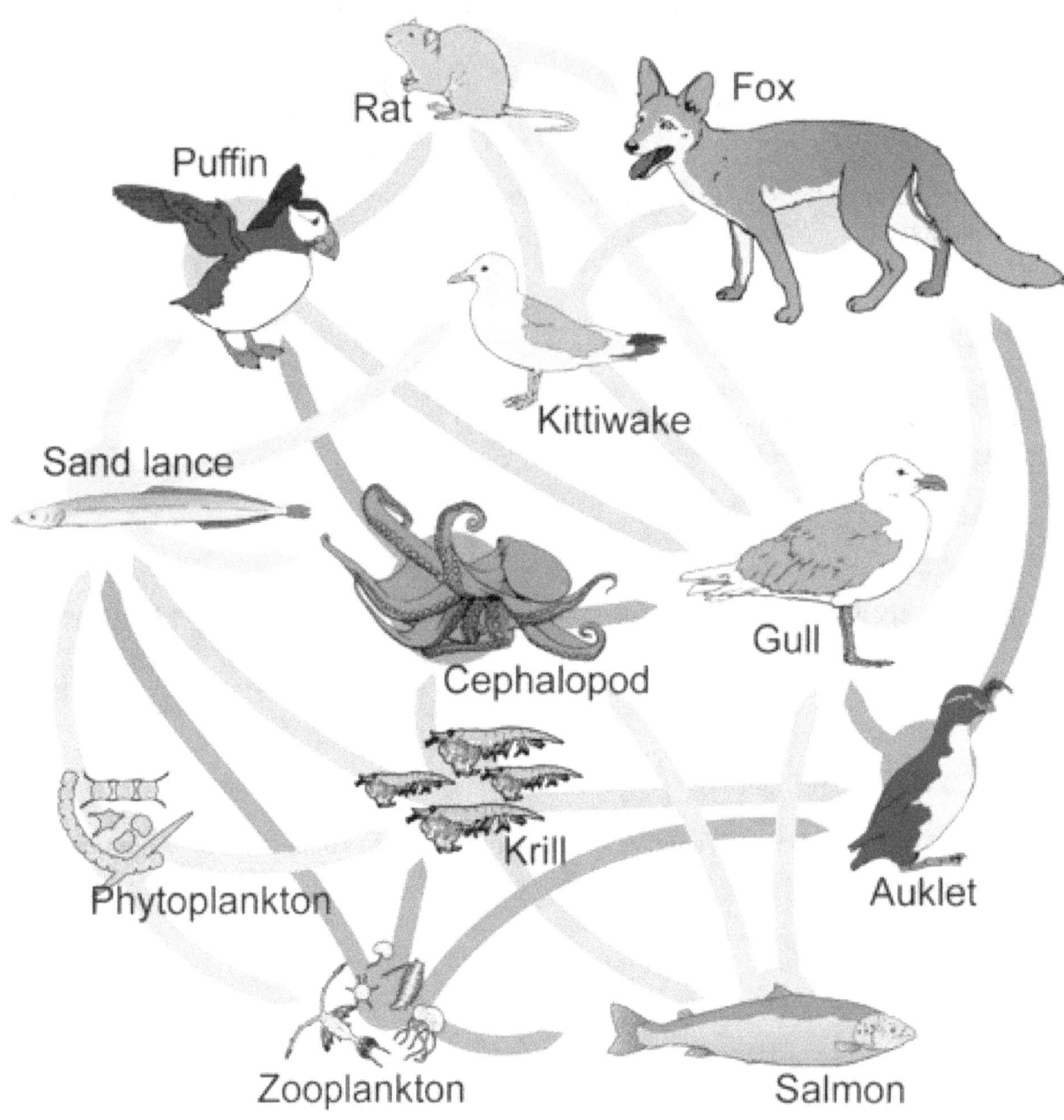

Picture from http://commons.wikimedia.org

Questions:

1) What would happen to the krill population if the Auklets were to go extinct?

2) What would happen to the bird populations if men in the area killed off the foxes?

3) What would happen to the populations of land animals if domesticated dogs were left in the area to breed?

4) What would happen to the populations if chemicals were released into the water that killed all the phytoplankton?

5) This community is in a cold coastal marine ecosystem. What would happen to each population if climate change made it warmer and brought more fish to the area?

6) What would happen to the bird populations if the fishing industry were to overfish the area?

    a. How would that affect the zooplankton population?

    b. How would that affect the fox population?

7) What would happen to the fish populations if whales were introduced to this community and ate all the krill?

    a. What would that do to the land animal populations?

Making a Food Web

Directions:

You will need large **butcher paper**, **scissors**, **glue**, **colored pencils**, a **ruler**, and a **meter stick**. **Looking at the materials and lab we will be using, what are the safety precautions we should take to protect ourselves and materials during the investigation?**

1) With colored pencils, mark each group a different color. Keep in mind that many organisms will have more than one color.
    a. Energy source – yellow
    b. Producers – green
    c. Herbivores – blue
    d. Carnivores – red
    e. Omnivores – orange
    f. Scavengers – purple
    g. Decomposers – brown

2) Use scissors to cut out the pictures. According to trophic levels, sort the pictures into groups: producers, herbivores, first-level carnivores, second-level carnivores, scavengers, and decomposers. Try to put the organisms near their food and predators (producers near the sun, insects in one area, and rodents in another). On some butcher paper, spread the organisms apart and give space to draw your arrows to see who eats who. Use arrows to point away from the food source and point to who is doing the eating.

3) Look at your food web; organisms with a star on them were sprayed with an insecticide DDT or had eaten an organism with it. If animals eat the organism that has been sprayed, they take in the poison. The organisms may not die, but the poison builds up in the organs of its body. Because predators eat more food that may be affected by the poison, more poison is concentrated in the higher-level consumers. **Place a red square** around the organisms in the food web that might get some poisonous DDT into their bodies from their food.

Toad **Consumer** Small Insects	Lizard **Consumer** Insects & Spiders	Squirrel **Consumer** Leaves & Seeds	Garden Snake **Consumer** Small animals
Fox **Consumer** Fruit & Small Animals	Skunk **Consumer** Insects, Small Rodents, & berries	Rattlesnake **Consumer** Small Animals	Jay Bird **Consumer** Insects, Seeds, & Spiders
Wolf **Consumer** Small & Large Animals, & Berries	Pheasant **Consumer** Plants & Insects	Rabbit **Consumer** Plant Leaves	Eagle **Consumer** Small Animals
Mouse **Consumer** Plants, Seeds, & Fruit	Beetles **Consumer** Plants	Spiders **Consumer** Insects	Ants **Consumer/Scavenger** Plants & Dead Animals
Butterfly **Consumer** Plant Nectar	Worms and Grubs **Consumer** Plants and Soil	Grass **Producer**	Oak Tree **Producer**
Cherry Tree **Producer**	Blackberry Bush **Producer**	Bacteria-Fungi-Molds **Decomposers** Dead Plants & Animals	Sun **Solar Energy** Source of Energy for Life

This page will be cut from the previous page.

Population Count

Directions:
Use **masking tape** to make a grid on the floor in 1-foot squares; enough for each person in the room. Randomly place a pre-counted amount of **beads** in the grid (the number is only known to the teacher). **Looking at the material and lab we will be using, what are the safety precautions we should take to protect ourselves and materials during the investigation?**

1) <u>Population Grid</u>: Each person will get assigned a square in the grid and count the number of beads in that grid. Put this number in Data Table 1 for one square.
2) Multiply that by how many grids there are to find the estimated population for the whole area studied. Put this data in the estimated population for one square in Data Table 1.
3) Randomly have students pair up and add their number of beads in their square together, divide by two. Put this in Data Table 1 for two squares.
4) Multiply by how many squares are on the floor. Put this number in Data Table 1 for the estimated population with two squares.

Data Table 1

	1 Square	2 Squares
Number of beads in square		
Estimated Population		

Questions:
1) Which (one square or two squares) had a closer estimated population to the real population (it does not always work but happens most of the time)?

2) Which (one square or two squares) had a closer estimated population to the real population for most people in the class?

3) Did anyone get the actual population in their estimate?

4) Why do you think it happened that way?

5) Why would we care about how many individuals there are in a population? Give as many different examples as you can.

6) What could be some sources of error in this investigation?

There are some other ways populations can be estimated talked about below:

1) Road Kill Count: Can count how many road kills see on the road to estimate the population of a type of animal. Easy to do in a car or truck. Good for looking at fluctuations in populations of animals like raccoons, dear, opossums, armadillos, and even stray dogs and cats. Good for smaller populations over a big area.

2) Population Survey: Walk through an area and count how many plants or animals you are looking for that you see. Estimate the area you covered to find the population of an area or population density. Good for smaller populations of bigger organisms over a big area.

Population Growth Curves

Directions:

Draw an exponential growth curve on Graph 1, where the graph rises faster and faster as time goes on. Draw a logistic growth curve on Graph 2, where it starts growing exponentially and then levels out. Where it levels out is the **carrying capacity:** the number of individuals of a species that an ecosystem can support. Label the carrying capacity on Graph 2. Use the graphs to help you answer the questions under them as you discuss them with your class and teacher.

Graph 1

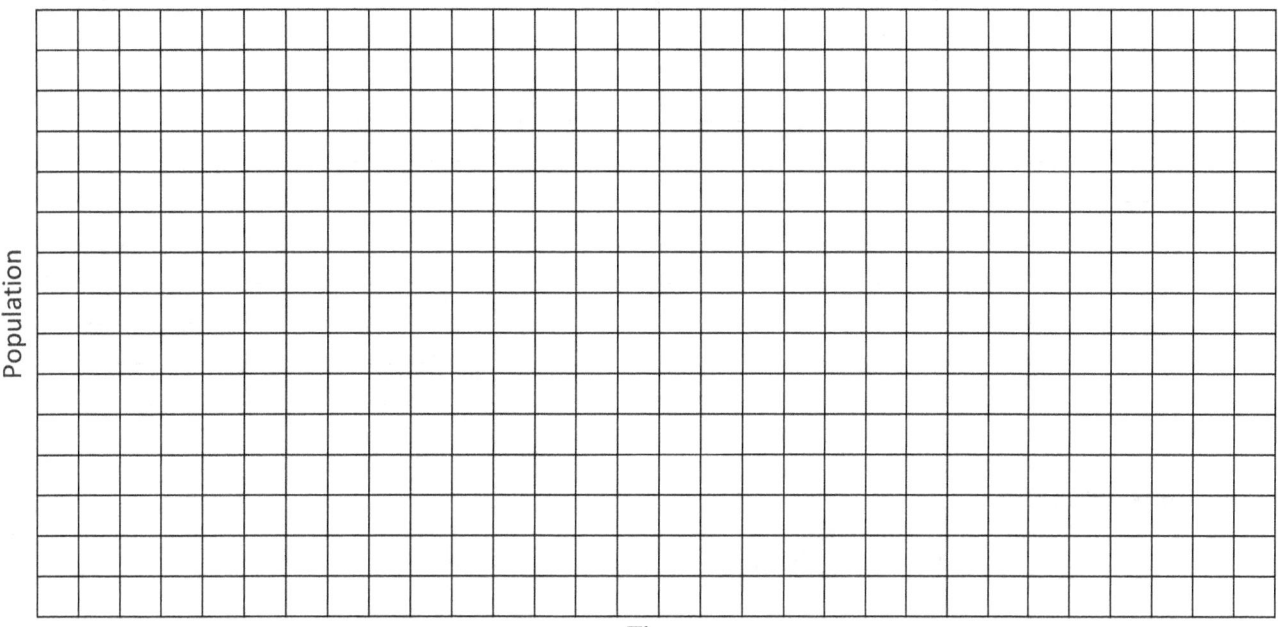

Time

Questions for Graph 1:

1) When a species has an exponential growth curve, can a population keep growing exponentially for a long time?

2) What are the two options that can happen to a population going through exponential growth?

3) When would we see exponential growth in a population of a species?

Graph 2

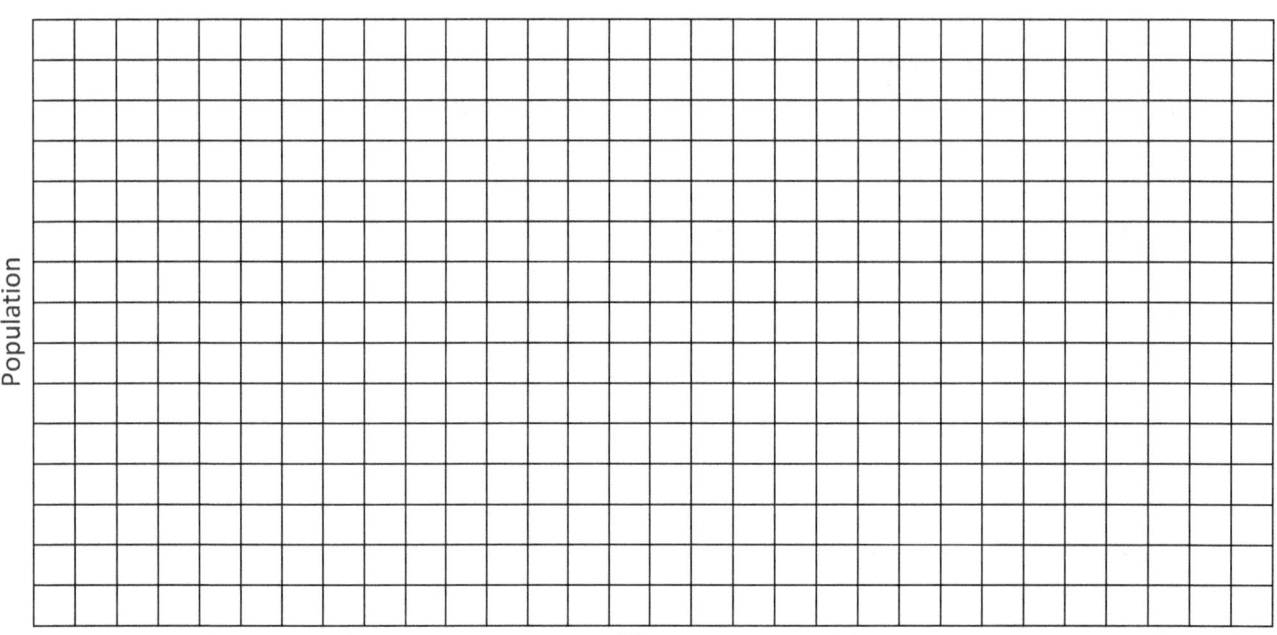

Questions for Graph 2:

1) What are some things that will raise the carrying capacity of a species?

2) What are some things that will lower the carrying capacity of a species?

3) What has to happen when a population goes past its carrying capacity? Explain why.

Causes of Invasive Species

1) Give three examples of how climate change is causing invasive species.

   a.

   b.

   c.

2) Give three examples of how humans introduce invasive species.

   a.

   b.

   c.

3) Give three examples of when humans are the invasive species.

   a.

   b.

   c.

Humans Changing Ecosystems

Directions:

Use the **internet** to research how ecosystems changed when humans added or removed these organisms from an ecosystem. Include which organisms were affected and how the food webs changed.

1) Humans moved to an uninhabited island, Mauritius, east of Madagascar, and brought pets.

2) Humans move into a new area and clear the land to build houses.

3) Humans release Burmese pythons Florida Everglades.

4) Humans release Asian carp into the Mississippi River to eat plants covering the river's bottom.

5) Humans plant crops on unused land.

6) Nutria were brought from South America to North America for the fur trade.

7) Wild boars were brought to the US for hunting and food.

8) Domesticated cats were bred and then released into the environment.

9) Humans chop down trees in the rainforest to make a field to raise cattle in Brazil.

10) In 1890 and 1891, European starlings were introduced to Central Park in New York City.

Biome Research Report

Have each student use the **internet** to fill out the following tables for a different biome to research, so all biomes are covered in the class.

Biome	
Location	
Climate	
Temperature include degrees °F (°C)	
Precipitation amount and pattern	
Soil description	
Plants and other producers	
Animals	
Additional Notes • human impact • interesting facts • connections	

Keystone Animals in my Biome

Notes: Be sure to indicate if the species is common, an endangered species, indicator species, keystone species, or introduced species.

Animal: Range: if migratory give summer and winter range Habitat description: Diet: Other: including interactions with other species.	**Animal:** Range: if migratory give summer and winter range Habitat description: Diet: Other: including interactions with other species.	**Animal:** Range: if migratory give summer and winter range Habitat description: Diet: Other: including interactions with other species.
Animal: Range: if migratory give summer and winter range Habitat description: Diet: Other: including interactions with other species.	**Animal:** Range: if migratory give summer and winter range Habitat description: Diet: Other: including interactions with other species.	**Animal:** Range: if migratory give summer and winter range Habitat description: Diet: Other: including interactions with other species.

Keystone Plants in My Biome

Notes: Be sure to indicate if the species is common, an endangered species, or an introduced species. You may also want to note if it is an indicator species or a keystone species.

Plant: Description:	**Plant:** Description:	**Plant:** Description:
Habitat description:	Habitat description:	Habitat description:
Importance: including interactions with other species.	Importance: including interactions with other species.	Importance: including interactions with other species.
Plant: Description:	**Plant:** Description:	**Plant:** Description:
Habitat description:	Habitat description:	Habitat description:
Importance: including interactions with other species.	Importance: including interactions with other species.	Importance: including interactions with other species.

Biomes Chart

Use the **internet** to fill in the chart for the characteristics of each biome.

Biome	General Description	Plant Adaptations	Animal Adaptations	Threats
Tundra				
Temperate Coniferous Forest (Boreal Forest/Taiga)				
Temperate Deciduous Forest				
Grasslands				

Biomes Chart Continued

Biome	General Description	Plant Adaptations	Animal Adaptations	Threats
Dessert				
Chaparral				
Tropical Savana				
Tropical Rain Forest				

Color in where each Biome is found on the Earth

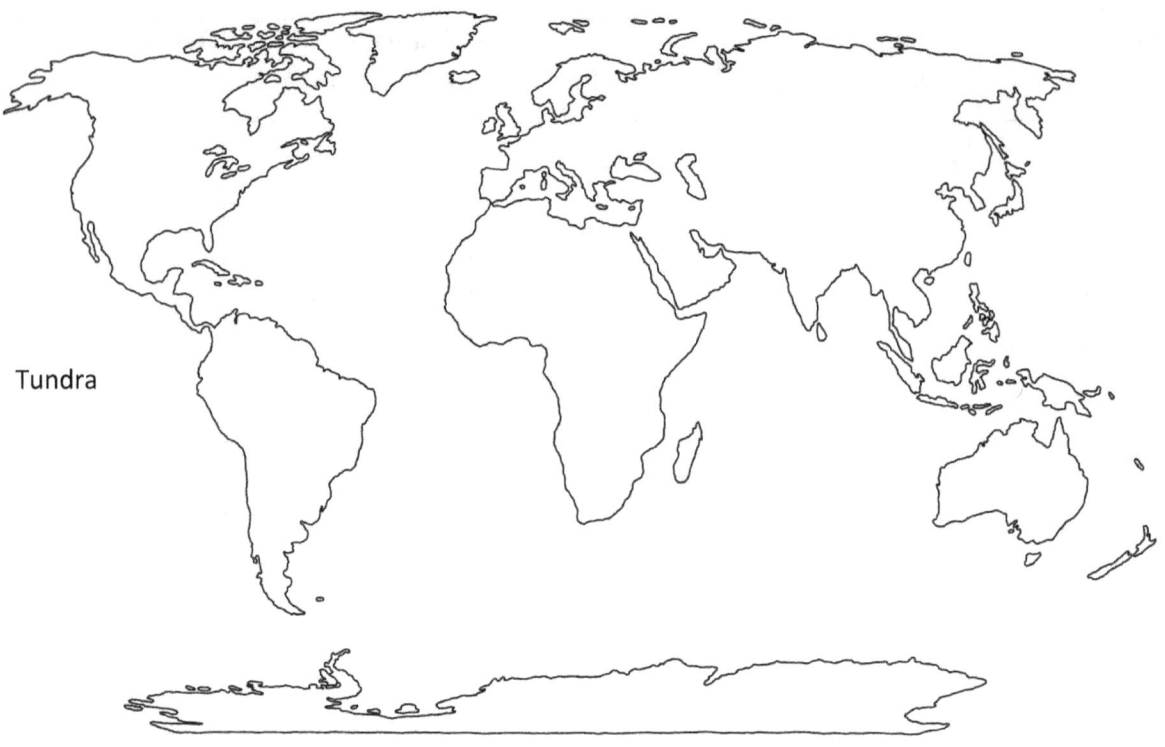

Tundra

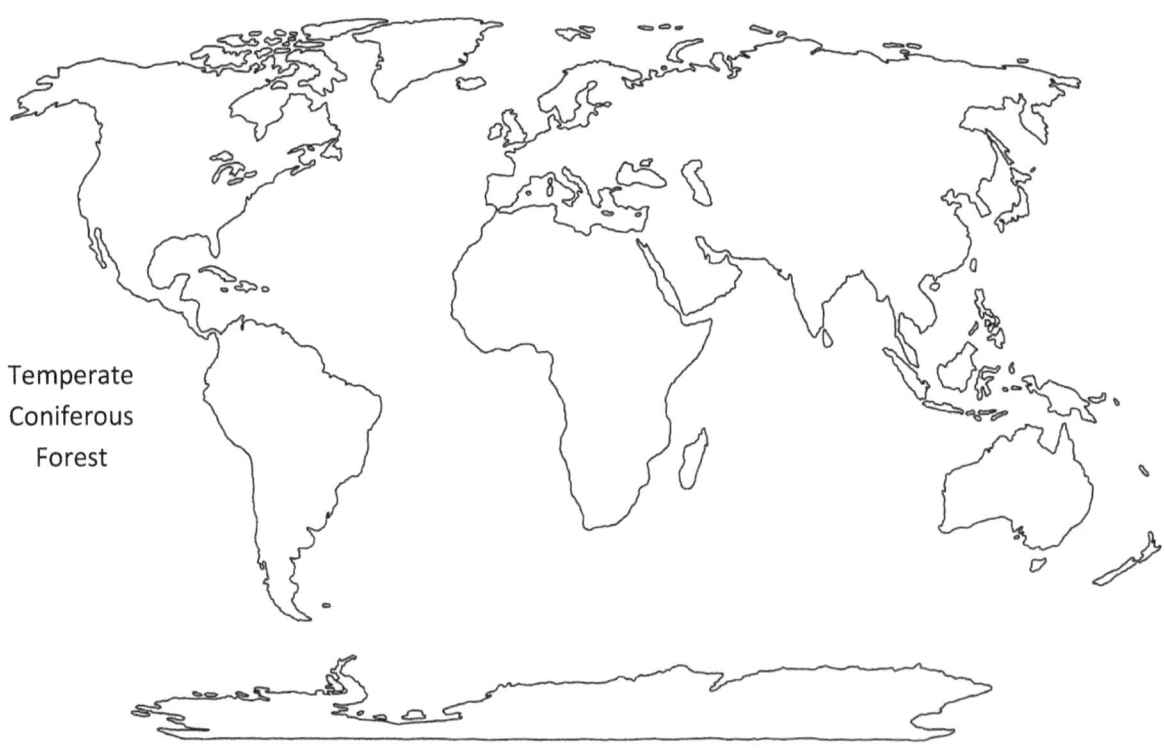

Temperate
Coniferous
Forest

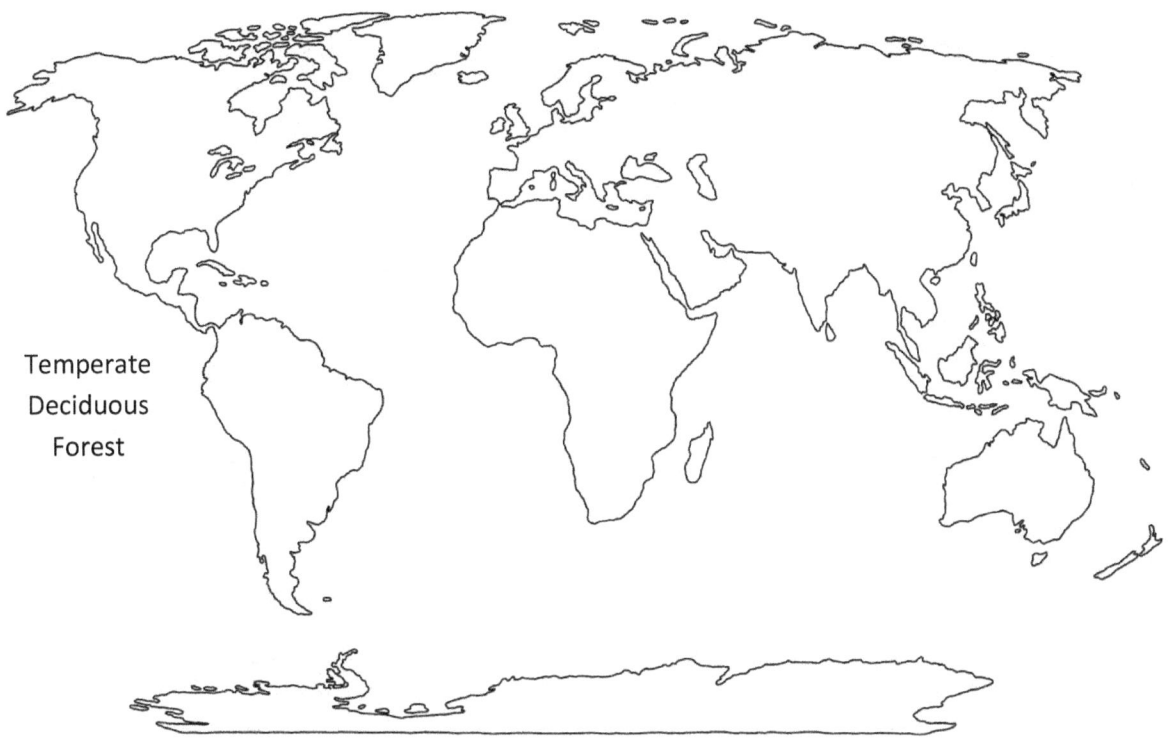

Temperate
Deciduous
Forest

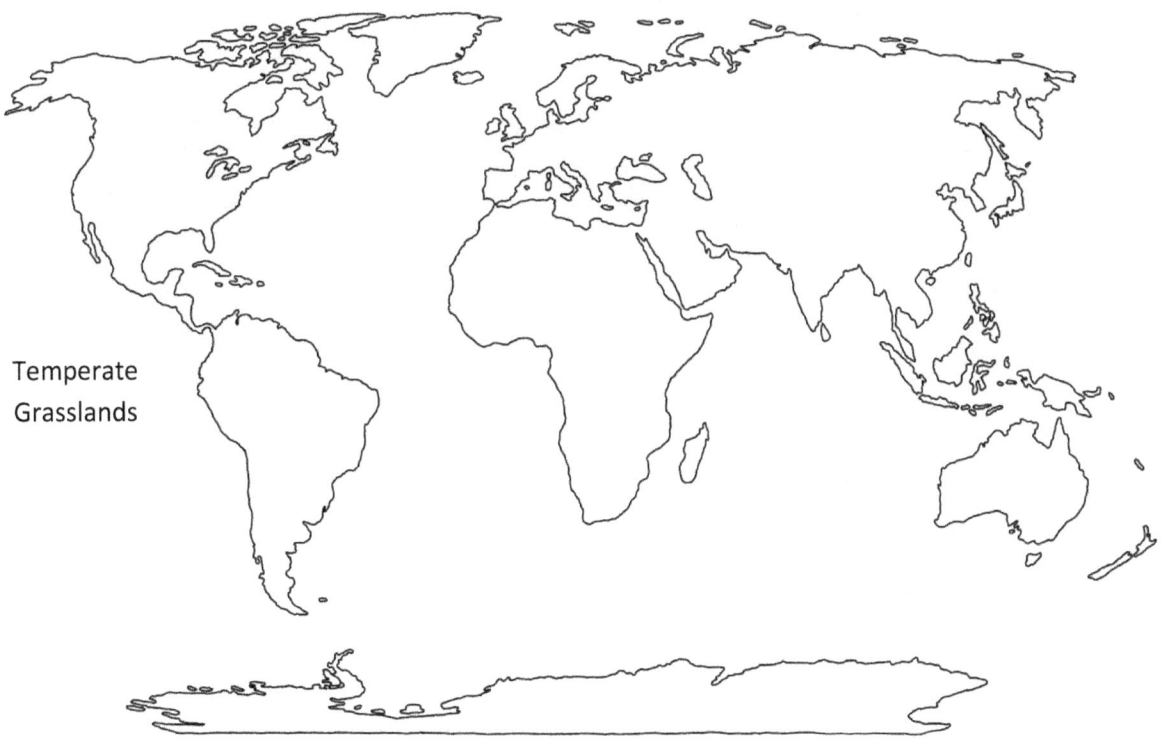

Temperate
Grasslands

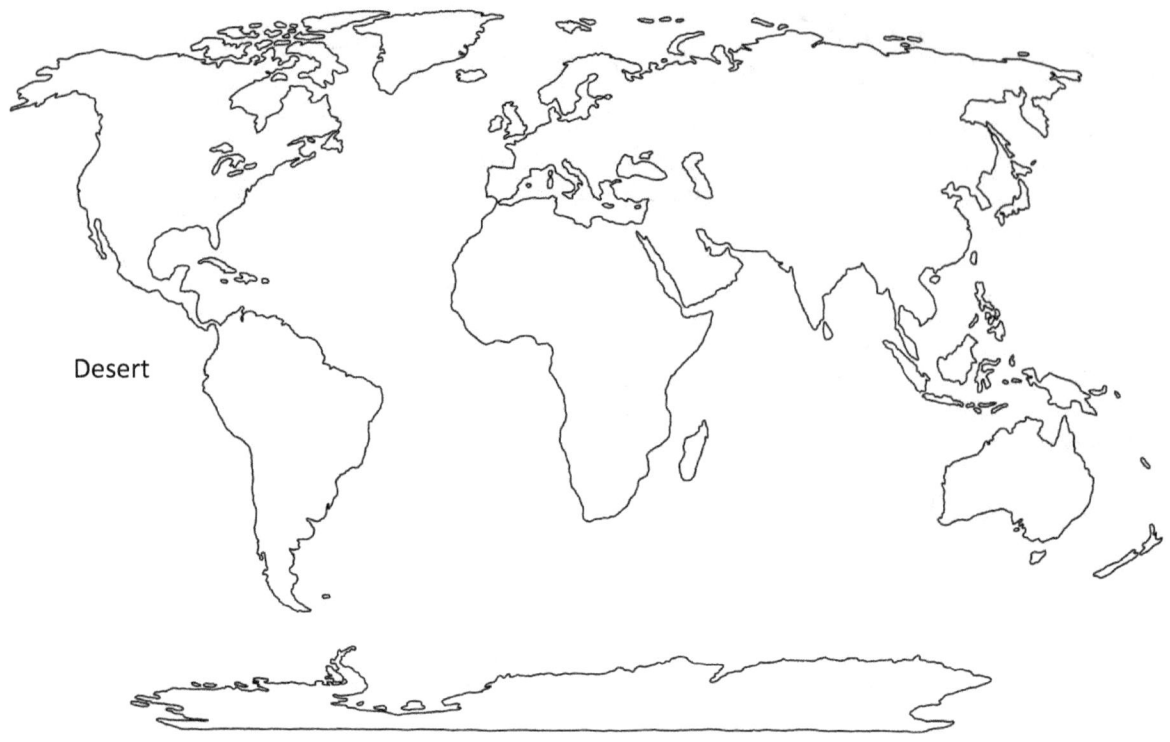

Desert

Chaparral

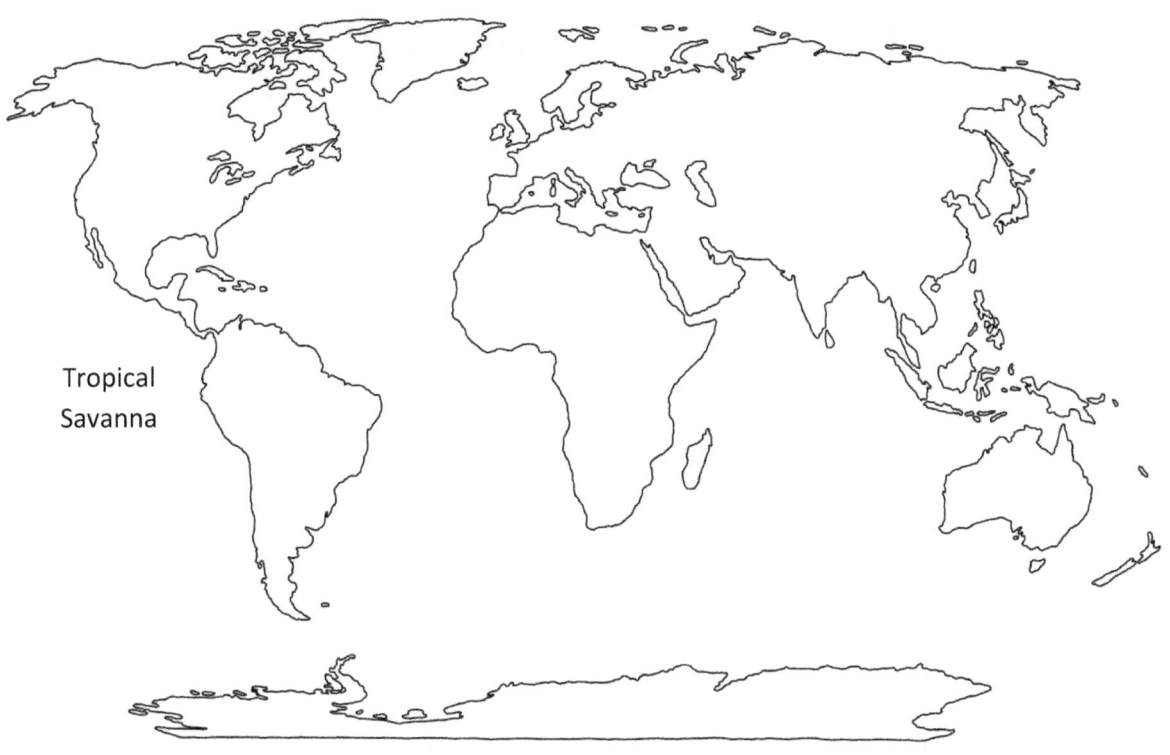

Tropical
Savanna

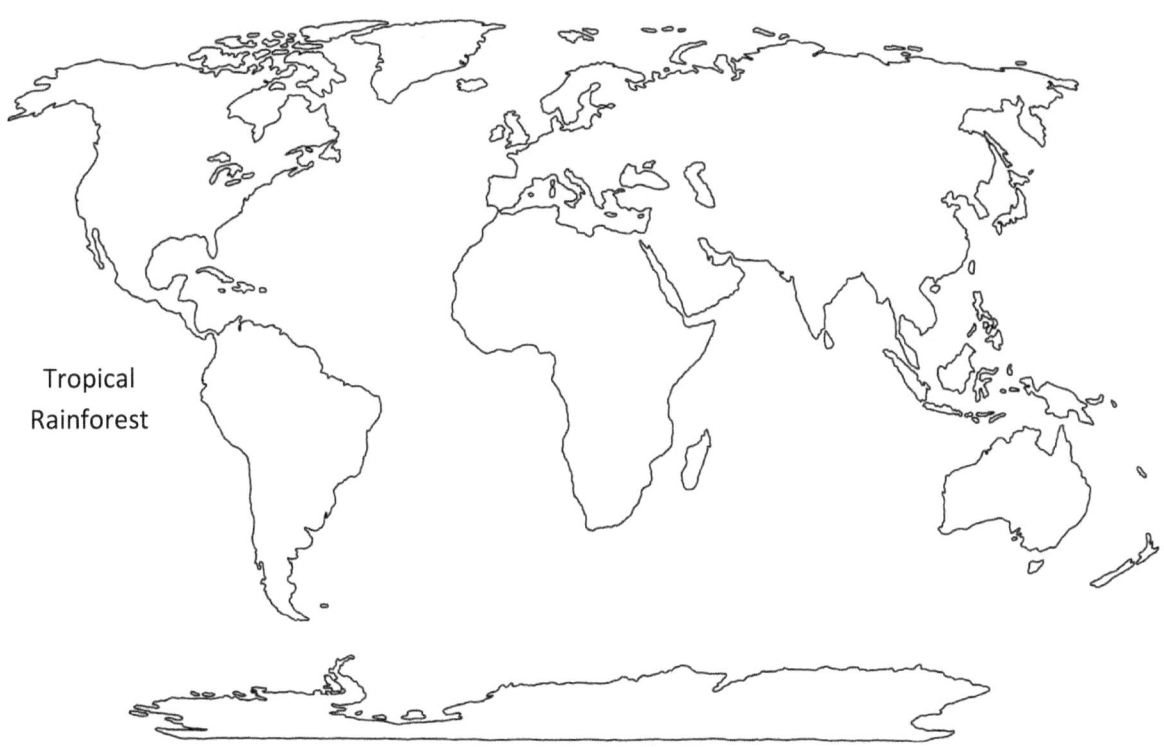

Tropical
Rainforest

Composition of the Atmosphere

Directions:

1) Use the **internet** to list the top 5 chemicals and their % in the atmosphere (include how much the water range can be).

2) Describe the Earth's first atmosphere and tell how it may have changed to what it is today. **Hint:** There were 3 atmospheres. Tell what the first atmosphere was composed of, then what happened to cause the second, then what happened to cause the one we have today.

The Greenhouse Effect

Directions:

You will need two **plastic tubs** (shoebox size) **painted black** on the inside, a **"Press'n Seal"** **sealing wrap**, two **temperature probes** attached to an **interface** connected to a **computer** with **Logger Pro**, and a **light source** like an incandescent lamp or some other lamp that gives off heat. **Looking at the materials and lab we will be using, what are the safety precautions we should take to protect ourselves and materials during the investigation?**

1) Make sure your tubs are both painted black on the inside, and a hole is poked through the end of each tub, big enough for a temperature probe to fit in snugly.
2) On the first tub, ensure the Press'n Seal is secured to the tub's opening, and the temperature probe is plugged into channel 1 of the interface.
3) The other tub needs to remain open, and the temperature probe plugged into channel 2.
4) In Logger Pro, open the folder Earth Science with Vernier and file #24 Greenhouse Effect.
5) Make sure your light source is equal distance from both tubs. Press "Collect" in the Logger Pro and turn on your lamp.
6) Monitor the time; when **5 minutes have passed, turn off the light**. Data will continue to be collected.
7) At the **10 minutes, turn the lamp back on**. Data collection will continue until 15 minutes. At 15 minutes, Data collection will stop.
8) Look at the data collected in the Logger Pro and fill in Data Table 1 for the initial temperature, the temperature at 5 minutes, the temperature at 10 minutes, and the temperature at 15 minutes.
9) Then subtract the temperatures between Probe 1 and Probe 2 to get the temperature differences at different times. Write this information on the right side of Data Table 1.

Data Table 1

	Probe 1 Greenhouse	Probe 2 Control	Temperature Difference
O Minute Temperature (°C)			
5 Minute Temperature (°C)			
10 Minute Temperature (°C)			
15 Minute Temperature (°C)			

Questions:

1) When the Lamp was on, which tub heated faster?

2) Give a possible explanation for your answer to number 1.

3) During the time when the lamp was off, which tub cooled faster?

4) Give a possible explanation for your answer for number 3.

5) How does this information show how temperatures could increase over time because of the greenhouse effect?

6) What are we doing to the atmosphere to increase greenhouse gases' effects to cause global warming and climate change?

7) How does this information show temperature could decrease over time without the greenhouse effect?

8) What do you think happened to the atmosphere to allow ice ages to happen on the Earth?

9) When do you think the temperature will get the coldest at night: when it is a cloudy night or when there are no clouds at night? Explain why.

10) The habitability of the Earth is a result of a delicate balance of the greenhouse effect. How/why is this statement true?

11) Explain why a closed automobile heats up in the sun.

12) Why do you not leave your child or pet in the car on warm days when the car is parked and turned off?

13) What could be some sources of error in this investigation?

14) Do you think the Greenhouse Effect is a hypothesis or a theory? Explain why.

Climate and Greenhouse Gases: Data Table

Directions:

Look at the Data Table below and build line graphs on the following pages showing the trends in temperature and carbon dioxide data. Then compare the trends in the graphs together, answering the questions that follow. This data was collected through ice core samples where the atmosphere was trapped inside the snow and then compressed into ice. When we melt the different layers of ice, we can accurately measure carbon dioxide levels from the past. Also, by measuring the number of Oxygen isotopes, we can get an accurate temperature of that time.

Data Table 1

Years Before Present (x 1000)	Local Temperature Change (°C)	Carbon Dioxide (ppm)
160	-9	190
150	-10	205
140	-10	240
130	-3	280
120	1	278
110	-4	240
100	-8	225
90	-5	230
80	-6	220
70	-8	250
60	-9	190
50	-7	220
40	-8	180
30	-7	225
20	-9	200
10	-2	260
0 (Year about 1850)	-0.5	280
0 (Year about 2002)	-	371

Graph 1 Local Temperature Change

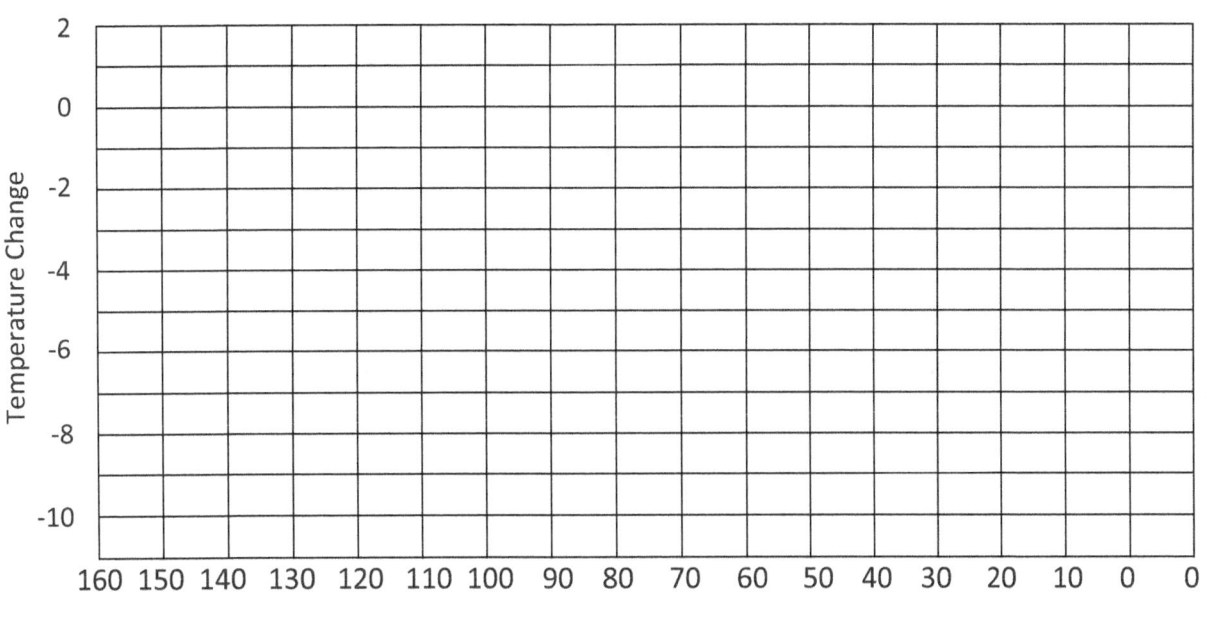

Age Thousands of Years Ago

Graph 2 Atmospheric Carbon Dioxide Levels over Time

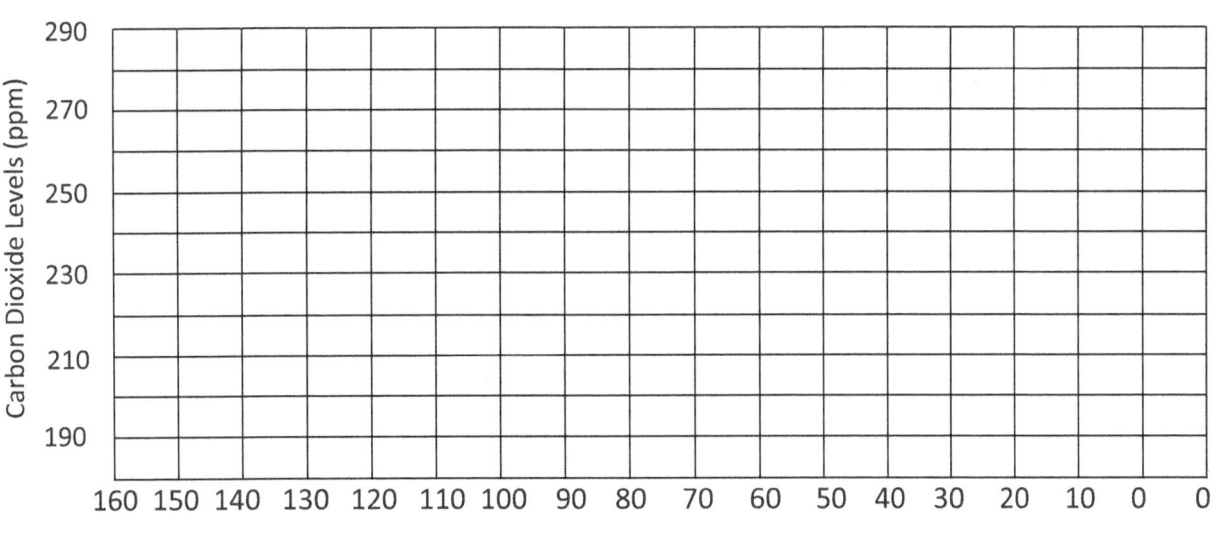

Age Thousands of Years Ago

Questions:

1) Do you see any similarities between the temperature graph and the Carbon Dioxide levels graph?

2) Which graph literally went off the chart and where?

3) What could this mean for the other graph in the near future?

4) Does this evidence show that Carbon Dioxide is causing the temperature change? Explain.

5) If the Earth is getting warmer, how might this affect our future climate and where people live?

6) How could this affect the evolution of life?

7) Can anything be done to prevent increasing temperatures? If so, what are they?

8) Do you think Global warming is a hypothesis or a theory? Explain why?

Carbon Dioxide and Population

Directions:

Graph the information from Data Table 1 onto Graph 1, then answer the questions below. Use the left Y-axis to create a line graph for the Population and the right Y-axis to create a line graph for the carbon dioxide emissions. Make sure you make a key showing which line on the graph is the population and which line is the carbon dioxide emissions.

Data Table 1

Year	Population (in millions)	Carbon Dioxide Emissions (in metric tons)
1750	790	11
1800	980	29
1850	1260	198
1900	1650	1,982
1950	2520	5,982
2000	6060	25,620

Graph 1

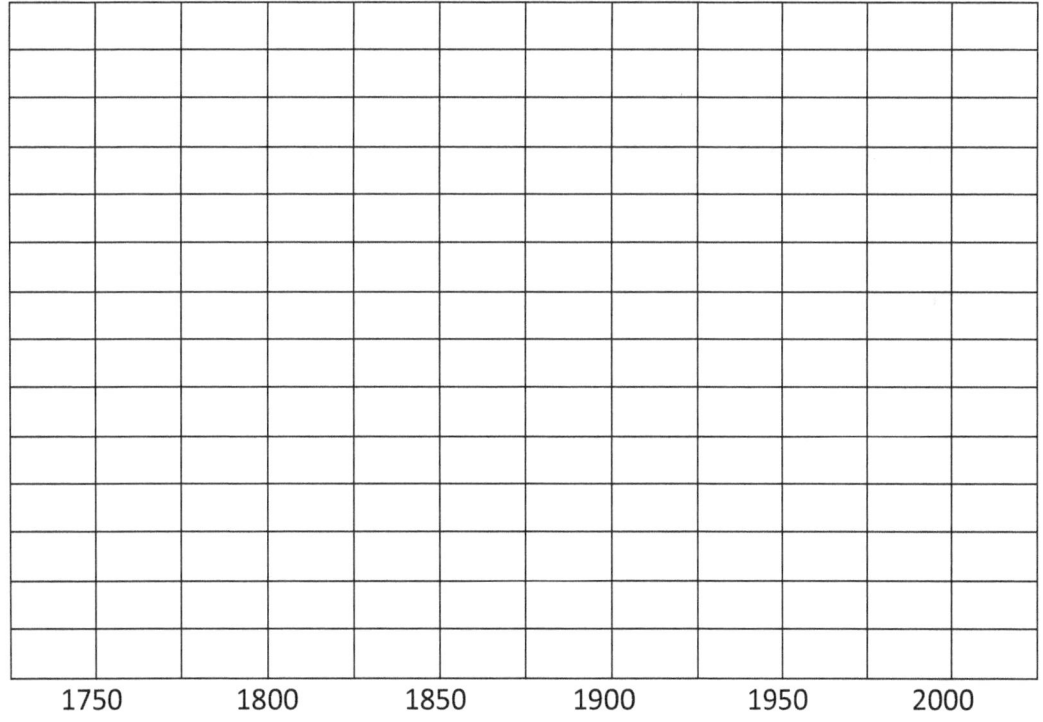

Year

Questions:

1) What happens to the carbon dioxide emissions as the population of humans rises?

2) Why do you think this happens?

3) How has the source of carbon dioxide emissions changed over the years?

4) Is there any way we can lower those emissions? If so, how?

5) What kind of culture change do we need if we are going to lower the carbon dioxide levels?

6) What is happening to the Earth because of the rising carbon dioxide levels?

7) How does this impact humans now?

8) How could this impact humans in the future if we do not change the trends in the data?

9) How could this impact the evolution of life in the future if the trends continue?

Climate Change

Directions:

Use the **internet** to research climate change and answer the questions below.

1) What is climate change?

2) What are the causes?

3) What is the impact of climate change on polar ice caps and glaciers?

4) How would this affect ocean currents?

5) How does it affect the surface temperatures of the Earth?

6) How is it impacting humans worldwide, and what can we expect in the future if this is happening?

7) What evidence do we have that it is happening?

8) What are the arguments against climate change?

9) Who are the people (occupations) studying climate change?

10) Do you think it is man-made or natural? Explain in detail why with evidence for your opinion.

How is Life Allowed on Earth?

Directions:

Use your **textbook** and **internet** as resources and what you have learned to explain the characteristics of the Earth that allow life to exist on it. Use this to answer the questions below:

1) What kind of sun/star do we need?

2) How far away does a planet need to be?

3) What effect does the rotation of a planet have on the ability for life?

4) How would a tilt of the axis affect the planet?

5) How does the moon help the Earth support life?

6) How does the atmosphere support life?

7) How does the magnetic field help the atmosphere on Earth?

8) What materials are needed for life to exist and thrive?

9) How would photosynthesis and aerobic respiration use these resources to help life thrive?

10) What can we learn about Venus and Mars to give us clues about the Earth?

11) What is the Goldilocks Zone, and how does that help life?

12) What things in the universe could destroy all life on Earth?

13) What protections are we getting from our solar system?

14) How complicated is the structure of life?

15) What does it take to make one cell the basic unit of life?

16) After researching this, how fragile is life on Earth, and how likely do you think we will find life on other planets?

    a.   Would it look like life on Earth? Explain why.

Our Little Mountain

Before I was born, my parents used to go on hikes and have picnics on this small mountain that overlooks a pass between two large mountain ranges. It was filled with tall pine trees with a clearing on the top. One day, shortly after I was born, we drove out to their special spot and saw that bulldozers were getting ready to knock down and uproot the trees. My parents were upset to see this. My dad saw a trailer parked close by with a makeshift parking lot. He went inside to find out what was going on. When he came out, he had a piece of paper in his hand and a scary smile on his face. He told my mom that they were selling land plots and were going to be building a winding road and houses on this mountain. My dad had just purchased the 5-acre lot at the top, which was their special picnic spot. We lived in Golden, Colorado, and he thought it could be a great place to put a vacation home if they could ever afford it one day.

A month later, we came back to see how the construction progressed. All the trees and bulldozers were gone, and a gravel road wound back and forth up our little mountain. My parents had packed a picnic that we would eat on our land at the top. When we had our picnic, the wind kept blowing dirt up into the air, and clouds of dust kept getting in our food and eyes; it was not very enjoyable.

Three years later, after my sister was born and able to walk, we went back to look at our mountain again. There was the greenest grass you ever saw up and down the mountain. Two houses were built, one at the bottom and an A-frame part way up. The one at the bottom served as an office for selling the plots on the mountain. We drove up the mountain and had a great time running around in the grass on our 5 acres. A crow flew down and snatched the peanut butter sandwich right out of my sister's hand. Everyone laughed so hard except for her. Later we saw some deer walk out of the woods next to the land cleared on the mountain. I looked out and imagined John Wayne driving thousands of head of cattle through the pass our land overlooked. My dad got a promotion a few weeks later, and we moved to Wisconsin for his new job.

Ten years later, my dad thought it would be fun for us to go on a family vacation, as he saw in a movie. He wanted to wind around America and eventually see our mountain. When we got there, we noticed lots of bushes and small trees were all over the mountain. The house at the bottom was abandoned. My dad did some research while we were there and found out we were only one of two that had purchased any plots, and the developer went bankrupt, and the mountain was abandoned. We went to the top of the mountain and picked berries growing on the bushes there. We saw a few rabbits and a small red fox running down the mountain across the dirt road just before leaving. I thought this could be a great place to put a house one day.

Twenty years later, I was now married with three children. My wife and I thought it would be fun to take our family skiing. I told her that my family owned some land on the top of a little mountain not far away from the lodge where we were staying. While on vacation, one afternoon, we took a drive to find the mountain. We found the little dirt road off the highway. As we drove up, we saw many leafy trees filling the mountain. We drove up the mountain until we reached the end of the road at the top. We walked through the forest until we reached a clearing at the top of the mountain, looking over the large valley pass that separated the mountain ranges on either side. I remembered the other times I had been there when I was younger. I wondered why my parents never built a house up there. There were lots of squirrels and birds. Pine tree seedlings seemed to be taking hold and growing around the clearing.

When I retired, and our oldest was visiting with our grandson, she asked about the land we visited in Colorado. My parents had died a few years before when COVID-19 hit, and I decided to look through the lockbox my father had given me before their passing. I noticed the deed on the land had been amended. The other owner moved off the mountain when the trees blocked their view of the valley below. My father was sold all the land surrounding the mountain. I now owned our little mountain. My publishing business took off, and I had enough money to build a house on the top of that clearing on the mountain. I hopped on a plane, went straight to our mountain, and saw that most of the leafy trees were gone. There were lots of small and medium-sized pine trees covering the mountain. I hired a construction company and built a large two-story cabin with big picture windows looking over both sides of the mountain. Because it was so far away from the nearest town and it was on the top of the mountain, it took a few years to build. My wife and I moved there after our youngest son graduated college. Our house also became the rest of our family's vacation home; my parents dreamed about when I was just a baby.

I was 95 years old when I died. My last view of this Earth was the tall pine trees that were just like the ones I saw when I was a baby. The trees filled the whole mountain like they did when my parents went on their hikes and picnics on the top of the mountain. These trees were the bottom of the frame of our bay window in the front of our house. I do not know if I imagined it or not. But I thought I saw a herd of cattle passing through the valley between the two mountain ranges just before I died.

Directions: Read each section of the story and draw a picture of the story's description in each box. When done with the story and your drawings, you should have a good picture of how succession looks.	**Before I was Born**
A Month Later	**Three Years Later**
Ten Years Later	**Twenty Years Later**
When I Retired	**I was 95 Years Old**

Questions:

1) What happened to the biodiversity when the land developers knocked down the forest?

2) During the story, when did the mountain have its greatest diversity? Explain why.

3) What happened to the plant populations when the forest was knocked down?

    a. Were there any homes for the animals there?

    b. What happened to the animal populations?

4) When did the grass populations increase?

5) When the trees started growing, what happened to the population of the grasses?

6) What happened to the kinds of plants during the story?

7) What happened to the kinds of animals during the story?

8) Who was the invasive species in the story, and how did they affect the ecosystem populations when they arrived?

9) From this story, can life find a way to recover from change? Explain your answer.

10) Who do you think will be outlived, humans or the rest of life? Explain.

Primary or Secondary Succession?

Directions:

Go outside, and find five examples of each.

1) Primary Succession takes place when rock is taking in life and eroding it away to make soil.

    a.

    b.

    c.

    d.

    e.

2) Secondary Succession occurs when something interrupts the ecosystem, reestablishing itself while keeping the soil.

    a.

    b.

    c.

    d.

    e.

Natural and Manmade Disasters

Directions:

Use the **internet** to fill in the following charts while researching natural and manmade disasters.

Natural Disasters

Name	What it does	Disasters that Cause it	Disasters it can Cause	Environmental Impacts	Impacts on Humans	Short Term Effects	Long Term Effects
Hurricane							
Flood							
Tornado							
Earthquake							
Volcanic Eruption							
Tsunami							
Mudslide/Avalanch							
Drought							
Wildfire							
Deforestation							

Manmade Disasters

Name	What it does	Disasters that Cause it	Disasters it can Cause	Environmental Impacts	Impacts on Humans	Short Term Effects	Long Term Effects
Oil Spill							
War							
Plane Crash							
Nuclear Reactor Meltdown							
Nuclear Bomb							
Mining Accident							
Municipal Development							
Fires							
Global Warming							
Fracking							

Virtual Investigations that go with the Ecology

ExploreLearning.com

Forest Ecosystem Gizmo

Rabbit Population by Season Gizmo

Food Chain Gizmo

Estimating Population Size Gizmo

Water Pollution Gizmo

Water Cycle Gizmo

Carbon Cycle Gizmo

Porosity Gizmo

Prairie Ecosystem Gizmo

Effect of Environment on New Life Form Gizmo

Temperature and Sex Determination Gizmo

Greenhouse Effect Gizmo

Energy Conversions Gizmo

Pond Ecosystem Gizmo

Ecosystems STEM Case Gizmo

Ecosystems Handbook Gizmo

Nitrogen Cycle STEM Case Gizmo

Nitrogen Cycle Handbook Gizmo

Corals Reefs 1 – Abiotic Factors Gizmo

Corals Reefs 2 – Biotic Factors Gizmo

Rock Cycle Gizmo

Seasons: Why do we have them? Gizmo

Summer and Winter Gizmo

Seasons in 3D Gizmo

Comparing Climates Gizmo

Observing Weather Gizmo

Ocean Tides Gizmo

Tides Gizmo

Coastal Winds and Clouds Gizmo

Weather Maps Gizmo

Relative Humidity Gizmo

pH Analysis Gizmo

Water Pollution Gizmo

Hurricane Motion Gizmo

Porosity Gizmo

Genetic Engineering Gizmo

GMOs and the Environment Gizmo

Energy Conversions Gizmo

Household Energy Usage Gizmo

Nuclear Decay Gizmo

Half-life Gizmo

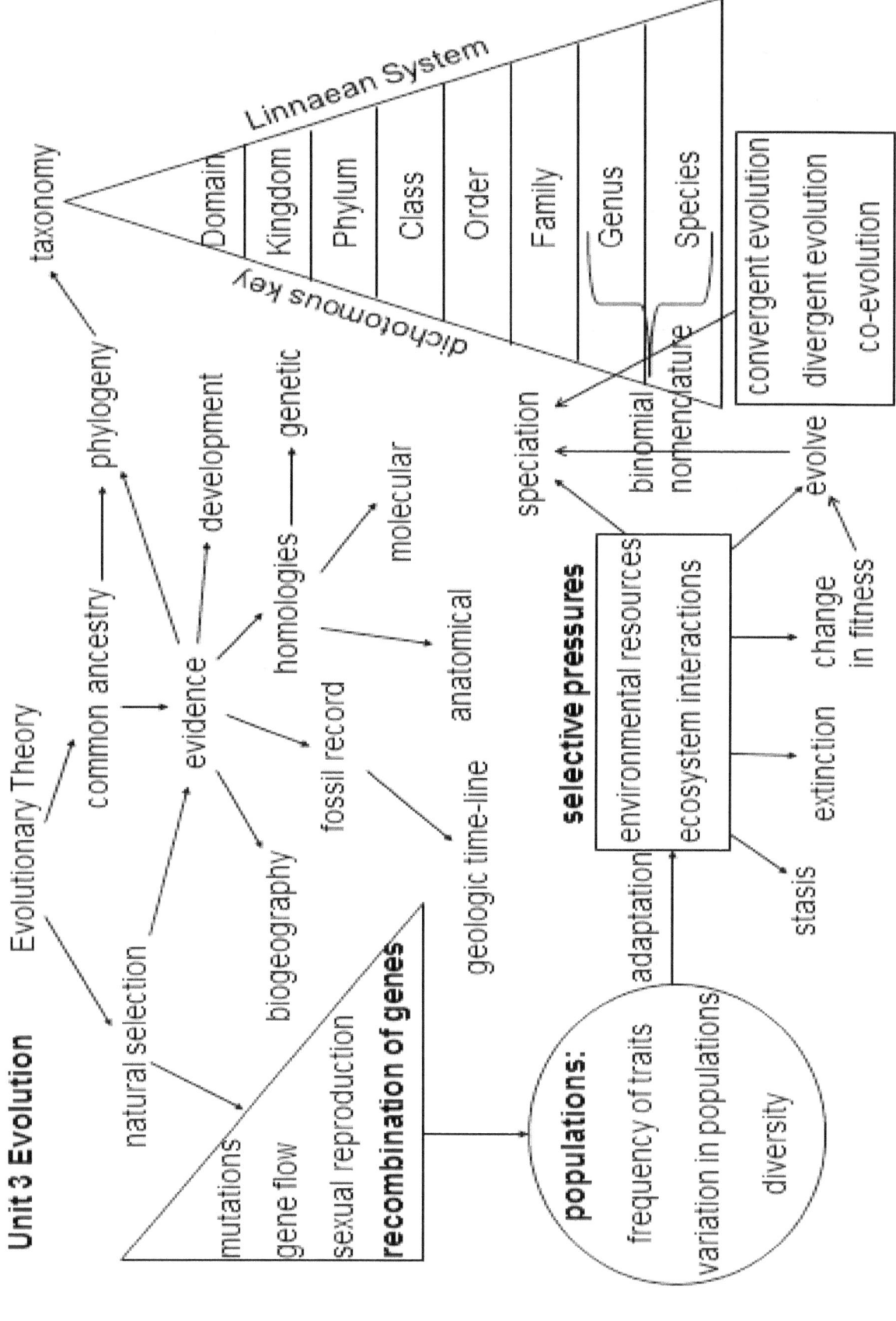

Analyzing Ancient Puzzles

Directions Part 1:

Look at the 100 million-year-old footprint pattern below taken from a riverbed. Follow the directions on the following pages to guide you through analyzing what may have happened so long ago.

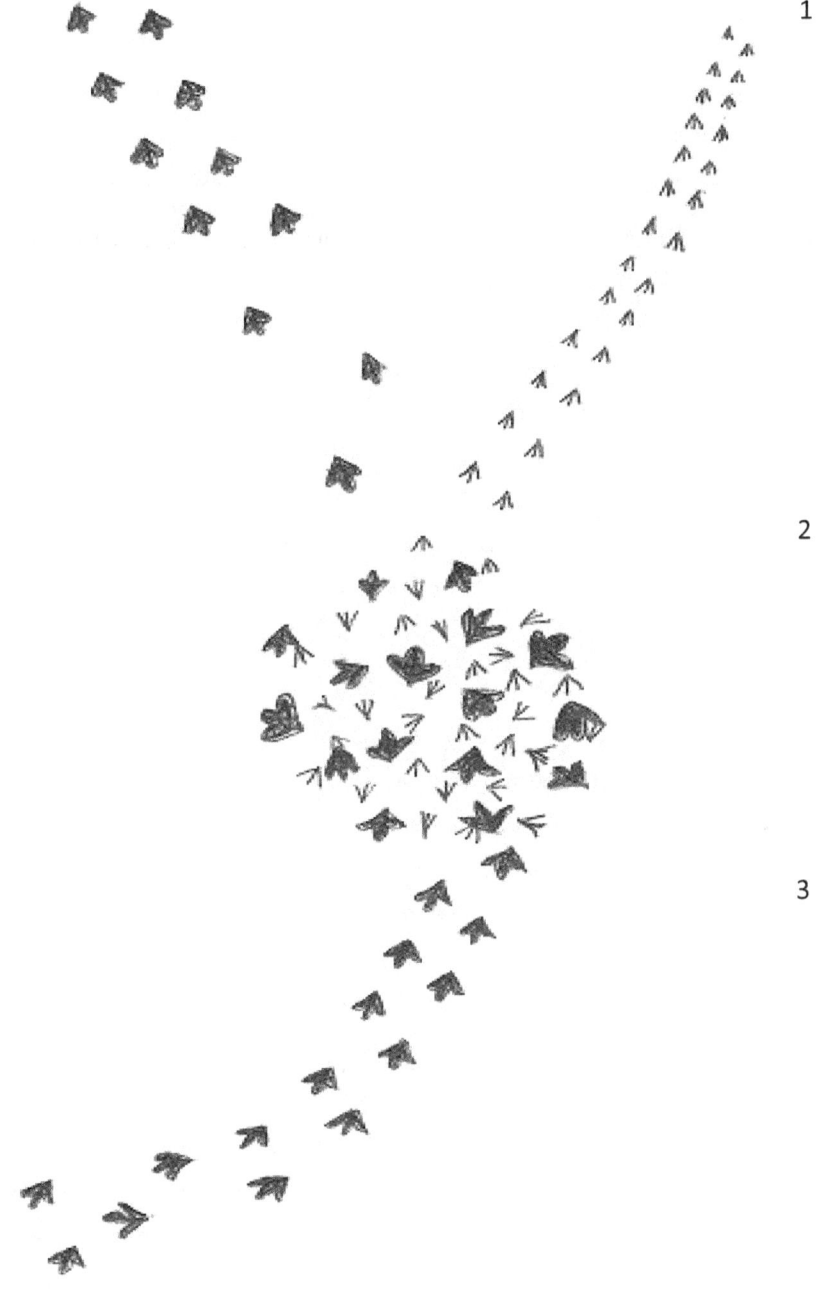

1) **Section 1 Observations**: With a piece of paper, cover the bottom 2/3 of the picture (by covering up to the number 2 and make three observations that you see in the picture on page 99 that no one can argue are not there.

    a.

    b.

    c.

2) **Section 2 observations:** Now cover only the bottom 1/3 (by covering the number 3) and make three more observations below.

    a.

    b.

    c.

3) **Section 1 and 2 inference:** Now explain what you think was going on to cause the footprints in these first two sections.

4) **Section 3 Observations:** Now uncover the entire picture and make three more observations that you see.

    a.

    b.

    c.

Section 1, 2, 3 Conclusion: Now tell a story about what you think caused all these footprints.

Questions Part 1:

1) Was it more accurate to describe what you think happened after part of the diagram or the whole thing? Explain why. (Think of it as a conversation, are you going to understand it by just hearing some of it or the whole thing?)

2) Is it possible that the animals who made the footprints never actually met each other when they were made? Tell how this could have happened.

Directions Part 2

Look at the picture below and answer the following questions that go with it.

Questions Part 2:

1) What animals do you think are making the footprints?

2) What do you think is causing the two solid lines?

3) What do you think was causing these patterns?

4) Did the animals that made these patterns have to interact with each other?

5) Using both diagrams, how do you know when an animal could be walking or running?

The Story of Life

Directions:

In some form or fashion, build a scale model of our history of life. Make the time equal to some standard length on a **meter stick**. Use the information below that we got from the fossil record to trace our lineage and build your Story of Life Time-line:

4.6 billion years ago, Earth formed

3.8 billion years ago, first life appeared

3.4 billion years ago, photosynthesis appeared

2.9 billion years ago, aerobic respiration appeared

2.0 billion years ago, protists appeared

1.0 billion years ago, primitive plants appeared

900 million years ago, fungi and simple animals appeared

700 million years ago, sexual reproduction appeared (the ability of life to mix genes speeding up evolution)

540 million years ago, oxygen levels rose, triggering the Cambrian explosion

530 million years ago, fish appeared

410 million years ago, amphibians appeared

360 million years ago, reptiles appeared

250 million years ago, temperatures rose from volcanic activity, and dinosaurs appeared

213 million years ago, mammals and birds appeared

200 million years ago, dinosaurs became the dominant creatures

65 million years ago, dinosaurs disappeared after an asteroid impact, triggering an ice age, and mammals became dominant

55 million years ago, primates appeared

7 million years ago, apes appeared

4 million years ago, hominids appeared

1 million to 100,000 years ago, <u>Homo sapiens</u> appeared

Questions:

1) How does this model show the Theory of Evolution?

       a. Why is it not a hypothesis?

       b. Why is this not a law?

2) When looking at this timeline, what do you think caused life to change on Earth?

3) How do you think the formation of Pangaea helped amphibians appear 410 million years ago?

4) If placental mammals did not exist in Australia but existed everywhere else on Earth, what does that tell us about how Pangaea started to break apart?

5) The fossil record shows that 99.9 % of all species that have ever lived on this Earth are now extinct. Why are they extinct, and why is the Earth not empty of life now?

Fossil Evidence of Relative Dating

Directions:

Use the diagram below modeling rock layers with a simple summary of different fossils. **Relative dating** is how scientists look at the history of life to see when things lived relative to each other. When we look at the rock layers, we find the oldest at the bottom and the youngest at the top. Use this diagram to explore this concept and answer the questions that follow.

Cenozoic Era	Hominids showed up Dominated by Mammals and Birds	A
Mesozoic Era	Dominated by Reptiles	B
Paleozoic Era	Dominated by Fish	C
Precambrian Era	Multicellular Invertebrates showed up	D
	Single Cell Eukaryotes showed up Single Cell Bacteria showed up	E
	No life found	F

Questions:

1) What kind of life was the first to show up?

2) Which kind of life was the last to show up?

3) 99.9% of all the species of life on this Earth are now extinct. Is this Earth empty of life?

   a. Could all the life live together at the same time?

b. So how could all this life have lived on this earth?

4) Which layer is the oldest?

5) Which layer is the youngest?

6) Mammals were around in the Mesozoic era; what do you think happened for them to be popular and dominate the Earth in the Cenozoic Era?

7) Describe the complexity of life as it formed through the history of the Earth.

8) Why are the older fossils found under younger fossils?

9) If different types of living things lived at different times, what does that say about the ecosystems?

a. What does that say about the abiotic factors that affected those ecosystems?

10) The one thing that never changes on this Earth is that everything changes. Will this Earth look the same as today in a billion years? Explain why.

Nuclear Decay Half-life of Pennies

Directions:

You will need a **plastic tub** (about the size of a shoebox) with a **lid** and 100 **pennies**. **Looking at the materials and lab we will be using, what are the safety precautions we should take to protect ourselves and materials during the investigation?**

1) A radioactive element's half-life is how long it takes for half the atoms to change into another element as they go through radioactive decay. Since pennies have two sides, almost half will land on heads when flipped, and almost half will land on tails.

2) In your tub, place all 100 pennies heads up. These will represent your radioactive isotopes.

3) Place the lid on your box, shake it and count to 10.

4) Lift the lid and take out all the pennies that have landed tails up. Count them and fill in Data Table 1. Subtract the number of tails from 100 to show the number still heads up.

5) Now only the pennies that are heads up are still in the tub. Repeat the procedure in #s 3-4 six more times unless all your pennies turned up tails before that.

6) Graph your data from Data Table 1 on Graph 1.

7) Your teacher will give each group a number and put your data into Data Table 2 for your group. Get the data for the other groups and put them in Data Table 2

8) Average each half-life for all the groups by adding their data and dividing by the number of groups.

9) Graph the averages you have for the class in Data Table 2 on Graph 2.

Data Table 1

Shaking Time (s)	# of Heads	# of Tails taken out
10		
20		
30		
40		
50		
60		
70		

Data Table 2

Groups	# of Heads at 0 (s)	# of Heads at 10 (s)	# of Heads at 20 (s)	# of Heads at 30 (s)	# of Heads at 40 (s)	# of Heads at 50 (s)	# of Heads at 60 (s)	# of Heads at 70 (s)
1	100							
2	100							
3	100							
4	100							
5	100							
6	100							
7	100							
8	100							
9	100							
10	100							
11	100							
12	100							
13	100							
14	100							
15	100							
Average	100							

Graph 1

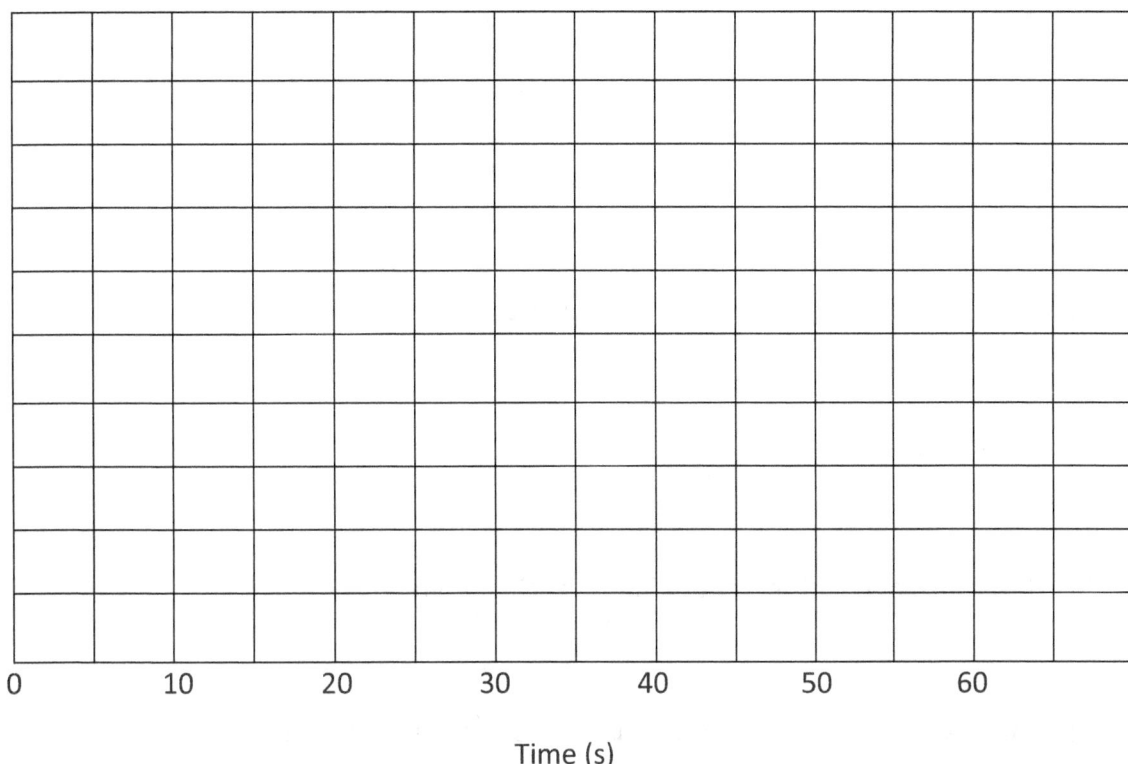

Time (s)

Graph 2

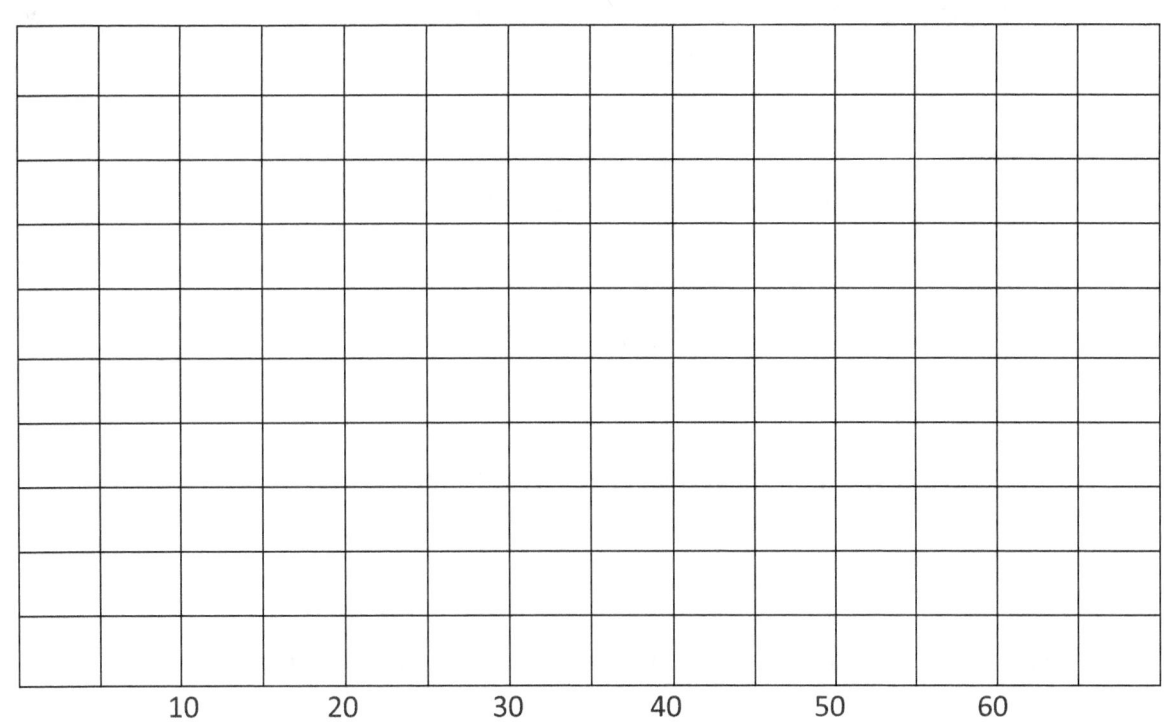

Time (s)

Questions:

1) What do the 10 seconds represent?

2) Which side of the coin was a stable isotope?

3) Which side of the coin was the unstable isotope?

4) What represented the radioactive decay?

5) How much of the atoms decay during each half-life?

6) How many half-lives until you should run out of pennies if all things go as expected?

7) If you started with 100 pennies and a half-life goes by every 10 seconds, how many pennies should there be after 40 seconds?

   a. How close was this to your group's results?

   b. How close was this to the class results?

   c. Why should the class results be closer to the expected?

8) How can the information in this lab be used to find the age of fossils?

Homologous Structures

Directions:

Use the drawings of animal forelimbs below to help you answer the questions that follow. Forelimbs are the appendages closest to the head in vertebrates. The sequence of bones for each of these, going from left to right, has a **humorous**, then **radius** and **ulna**, then small **carpal** bones, then skinny **metacarpal** bones, and finally, the even smaller **phalanges**. Use **colored pencils** to color-code each type of bone below.

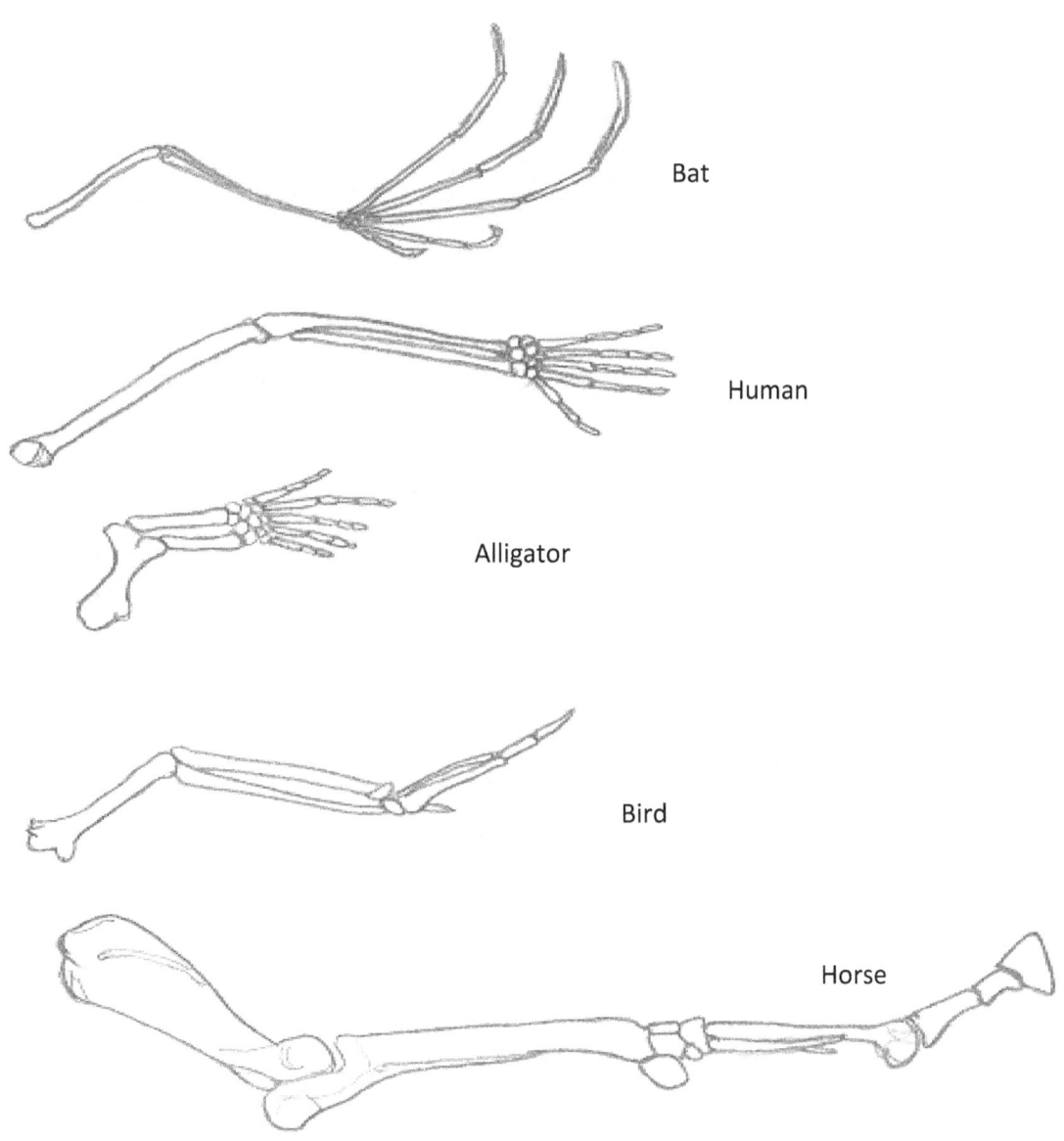

Bat

Human

Alligator

Bird

Horse

Questions:

1) What similarities do these animals have?

2) What differences do these animals have?

3) Contrast the phalanges (finger bones) of each of these animals. How are they different?

4) Do these animals use their bones for the same function? (Explain)

5) How can we tell that these animals have common ancestry?

6) How are these pictures evidence of evolution taking place?

Evidence in Embryonic Development

Part 1 Information:

All animals, when they reproduce sexually, go through these stages seen below. They start as a zygote and then divide cells until they form a blastula. After that, the first hole called a blastopore, forms. Vertebrates are deuterostomes, and their blastopore turns into an anus. Deutero- is a prefix that means second (our second hole forms first). Most invertebrates are protostomes, and their blastopore turns into a mouth. Proto- is a prefix that means first.

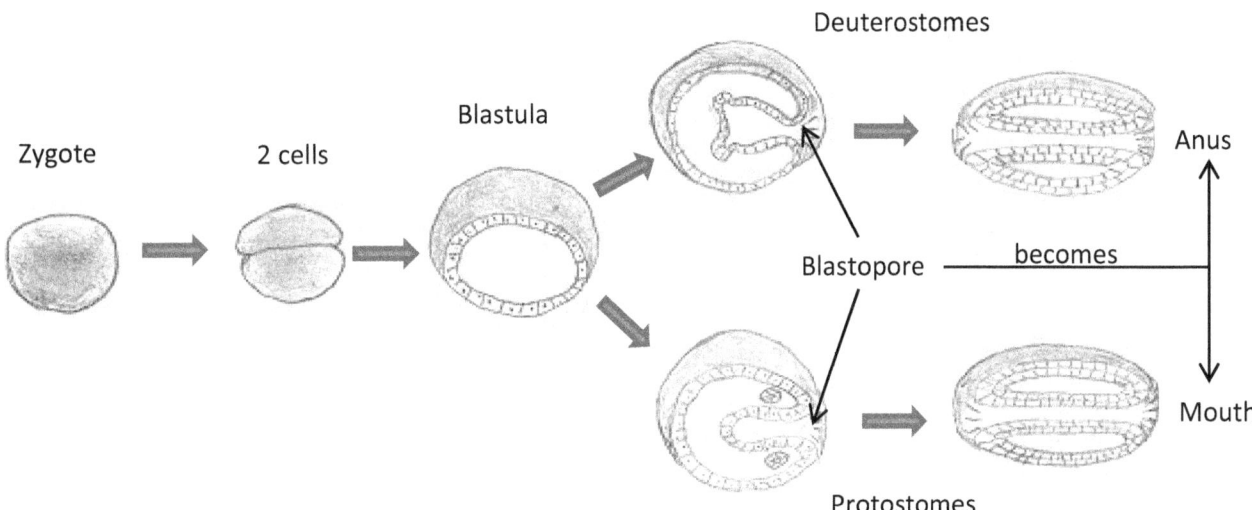

Part 1 Questions:

1) How does this picture show that vertebrates and invertebrates are related?

2) How does it show they are different?

3) How does this drawing show the history of their ancestry?

4) According to this drawing, which came first, worms or insects?

Part 2 Information:

As vertebrates further grow and develop, more characteristics appear, showing you what that organism is. Many of these organisms take more or less time than others to reach their different developmental stages, but they each go through the same general stages. At some time in their lives, all vertebrates have a dorsal notochord, pharyngeal gill slits, and a post-anal tail (including us, as you can see). Use the picture below of three embryonic stages of these six vertebrates to answer the following questions.

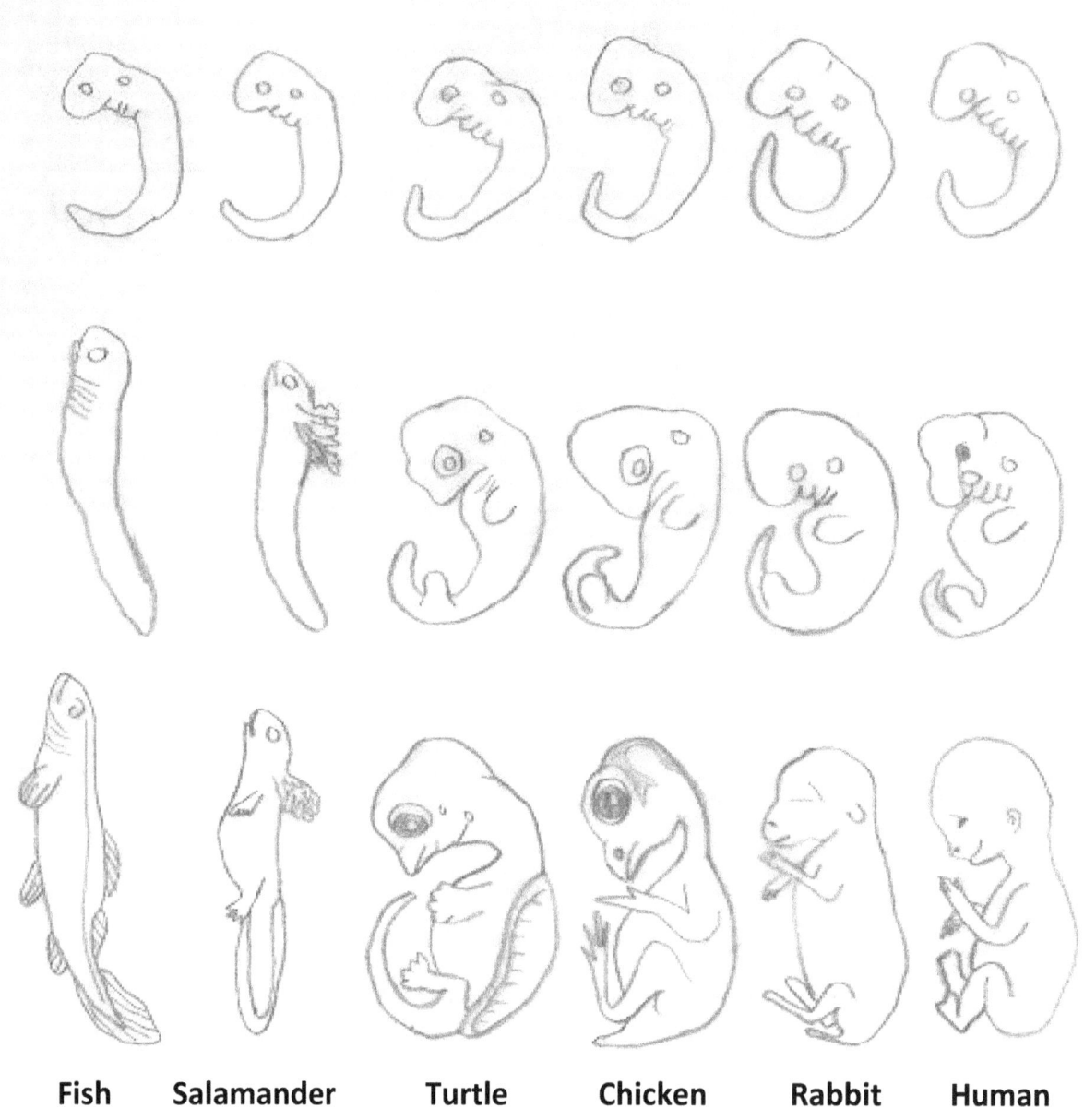

Fish   Salamander   Turtle   Chicken   Rabbit   Human

Part 2 Questions:

1) If we look only at the top row, can you tell the difference between any of these animals?

2) As you look at the diagram on page 114, which animal looks most closely related to the fish?

3) Which animal appears to be the most closely related to humans?

4) Which two animals here lay amniotic eggs?

5) How does their development show evidence that they are related?

6) How do these pictures show evidence of evolution?

7) Of all the pictures in this activity, at which stage do you think we would stop looking like a plant? (Explain)

Comparing Relatedness with Proteins

Directions:

Look at the sequence of amino acids in the chart below. The more similarities in the code, the closer the relatedness between organisms. The fewer similarities in the code, the less relatedness between the organisms. Genes that organisms share can change at a constant rate. Each amino acid is coded by a letter of the alphabet. Just like letters of the alphabet make meanings by building words and sentences, amino acids can be put together to make meaning by building proteins. Fifty positions are labeled for the human, turtle, shark, and fruit fly.

1) Find where there are differences between the humans and the other animals by circling or highlighting where the other animals are different from the humans. Write this number in Data Table 1 and calculate the percent difference.

G-Glycine A-Aspartic Acid V-Valine U-Glutamic Acid L-Lysine I-Isoleucine P-Phenylalanine M-Methionine C-Cysteine S-Serine B-Glutamine H-Histidine T-Threonine V-Valine D-Proline N-Asparagine E-Leucine F-Arginine K-Tyrosine O-Alanine Q-Tryptophan

| 1 | 5 | 10 | 15 | 20 | 25 | 30 | 35 | 40 | 45 | 50 |

Human GAVULGLL I PI MLCSBCHTVULGGLHLTGDNEHGEPGFLTGBODGKSKTO

Turtle GAVULGLL I PVBLCOBCHTVULGGLHLTGDNENGEI GFLTGBOUGPSKTU

Shark   GAVULGLLVPVBLCOBCHTVUNGGLHLTGDNESG EPGFLTGBOQGFSKTD

Fruit fly GAVULGLLEPVBFCOBCHTVUOGGLHLVGDNEHGEI GFLTGBOOGPOKTN

| 1 | 5 | 10 | 15 | 20 | 25 | 30 | 35 | 40 | 45 | 50 |

Data Table 1

Humans compared to:	Number of Differences	Percent Difference
Turtle		
Shark		
Fruit Fly		

The fossil record shows that humans diverged from reptiles about **235** million years ago, fish **420** million years ago, and arthropods **590** million years ago. Take the percent difference and calculate the percent change per million years using this information.

2) Take the percent difference and divide it by how long ago we last had a common ancestor with these organisms. Write this in Data Table 2

3) Calculate the Average percent change per million years by adding the three animals' data and dividing by 3. Write this data at the bottom of Data Table 2.

Data Table 2

Reptiles % change per million years	
Fish % change per million years	
Arthropods % change per million years	
Average % change per million years	

Questions:

1) If fungus last had a common ancestor with us 900 million years ago, how much change should we see with this gene in a fungus? (Hint: reverse the calculations we just did)

2) If plants diverged from us 1500 million years ago, how much change should we see with plants?

3) Which of the organisms we looked at in this activity shared the most characteristics with us?

4) Which of the organisms we looked at in this activity shared the least amount of characteristics?

5) How does this show evidence of evolution?

Biodiversity in Ecosystems

Directions:

Use the **internet** to fill in the chart below to find and rank the biodiversity of each biome. Then answer the questions that follow.

Biome	General Description	Plant Diversity # of Species	Animal Diversity # of Species	Diversity Rank
Tundra				
Temperate Forest				
Desert				
Grasslands				
Tropical Rain Forest				

Questions:

1) Which of the Biomes in the chart showed the greatest biodiversity?

      a. Why do you think there is more life there?

2) Which of the biomes showed the least diversity?

      a. Why do you think there are fewer types of life there?

3) How does biodiversity show us an area's ability to support life?

4) If there is a change in the environment, which is more likely to survive the change, an ecosystem with high biodiversity or an ecosystem with low biodiversity? Explain why.

5) Does biodiversity increase or decrease around human populations?

     a. Why do you think this happens?

6) When humans grow crops, how many species are planted in one field?

     a. Does this increase or decrease biodiversity?

7) Where do you think there is greater Biodiversity in an energy pyramid (in the producers, herbivores, or carnivores)? Explain.

8) How does diversity help support an ecosystem?

9) Which ecosystem would be easier to destroy, one with high biodiversity or one with low biodiversity? Explain Why.

Variation Within a Population

Directions:

Each student will need ten **leaves** of any kind (must all be of the same species), ten **shelled nuts** or **seeds** (all of the same species) of any kind, and a **metric ruler**. The more students you have, the bigger your data set and the better results you will get. **Looking at the materials and lab we will be using, what are the safety precautions we should take to protect ourselves and materials during the investigation?**

1) Go out to a tree and randomly take ten leaves off the tree. When you take the leaf off, make sure you do not rip any part of the leaf. Measure the length of the longest part of the leaf in millimeters and write this data in Data Table 1.

2) Randomly take ten whole-shelled nuts or seeds out of a bag. Do not use the ones that are broken. Measure the length in millimeters and write this data in Data Table 1.

3) Find the measurement of the shortest leaf in the class and the longest leaf in the class. Then fill in the equal increments between those measurements to make 14 different groups. Then take a class count of how many leaves fit in each of those categories. Write this data in Data Table 2.

4) Find the shortest nut pod/seed and the longest nut pod/seed in the class. Then fill in the equal increments between those measurements to make 14 different groups. Then take a class count of how many nut pods fit in each of those categories. Write this data in Data Table 3.

5) Measure your pinky length from the crevice next to it (without stretching the webbing between your pinky and your ring finger) to the tip. Do not count the fingernail. You might want to total your data for the whole day to have enough numbers to show good data. Have your teacher keep your measurement on their roster. What is the length of your pinky?

6) Find the measurement of the shortest pinky of all the classes and the longest pinky of all the classes. Then fill in the equal increments between those measurements to make up 13 different groups. Then take a class count of how many pinkies fit in each of those categories. Write this data in Data Table 4. Then fill in the pinky data and let the students copy the all-day count the next day.

7) Make a graph of the class counts of tree leaf data on Graph 1.

8) Make a graph of nut pod/seed length data on Graph 2.

9) Make a graph of the pinky data for the whole day for Graph 3.

Data Table 1

	1	2	3	4	5	6	7	8	9	10
Leaf Length (mm)										
Nut Pod length (mm)										

Data Table 2

Leaf Length (mm)														
Class Count														

Data Table 3

Nut Pod Length (mm)														
Class Count														

Data Table 4

Length of Pinky (mm)														
Class Count														
All-day Count														

Graph 1

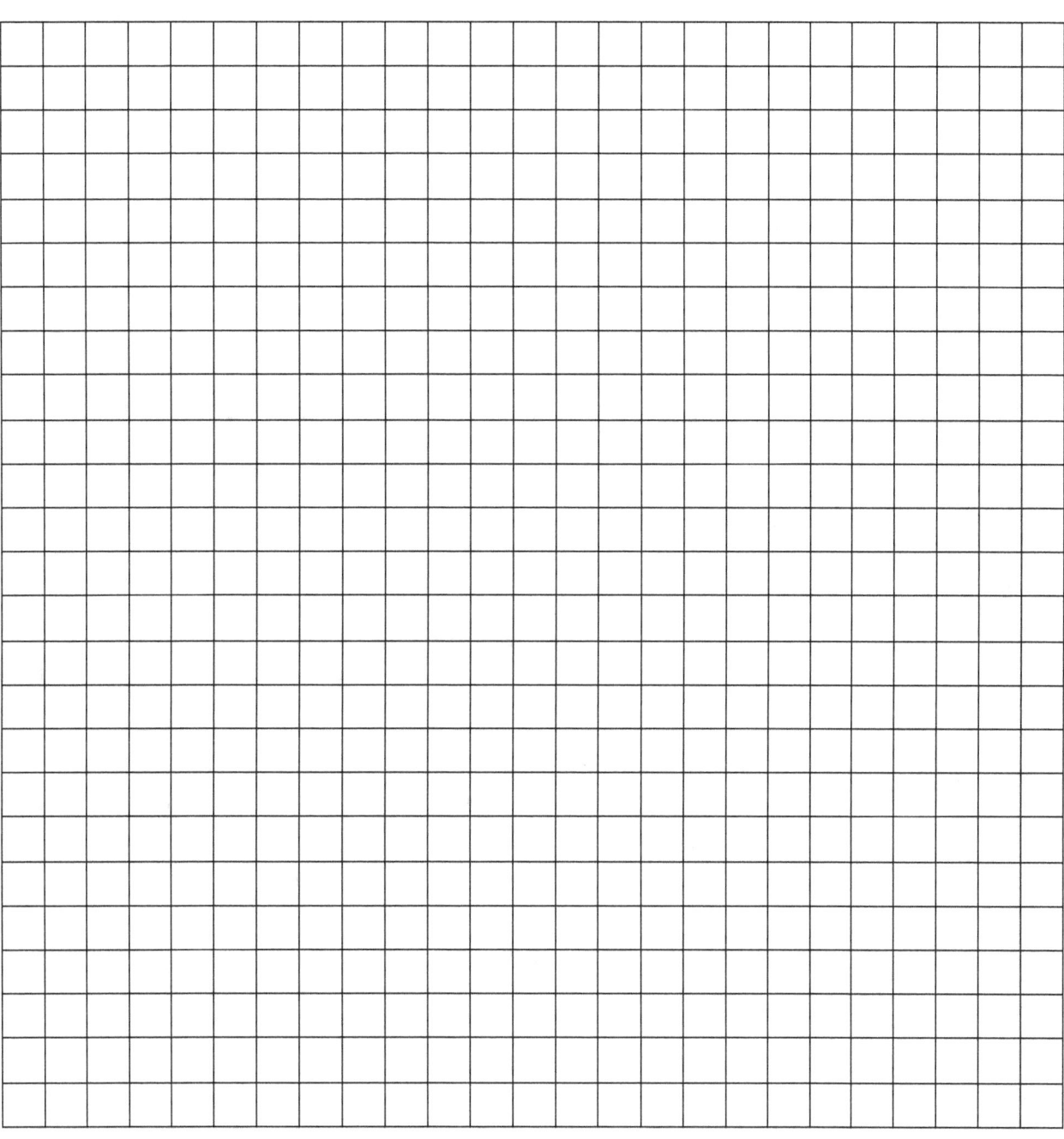

Graph 2

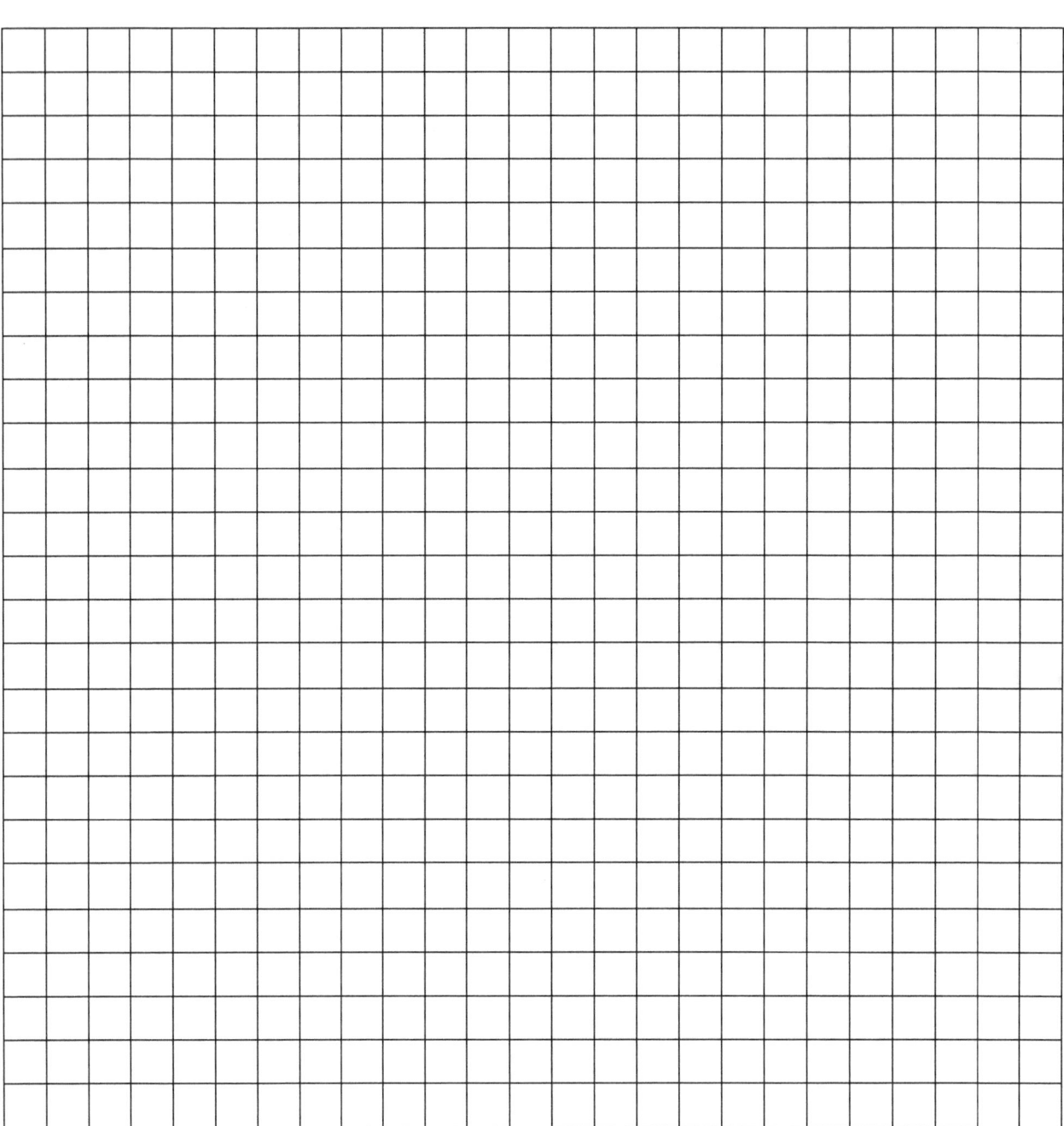

Graph 3

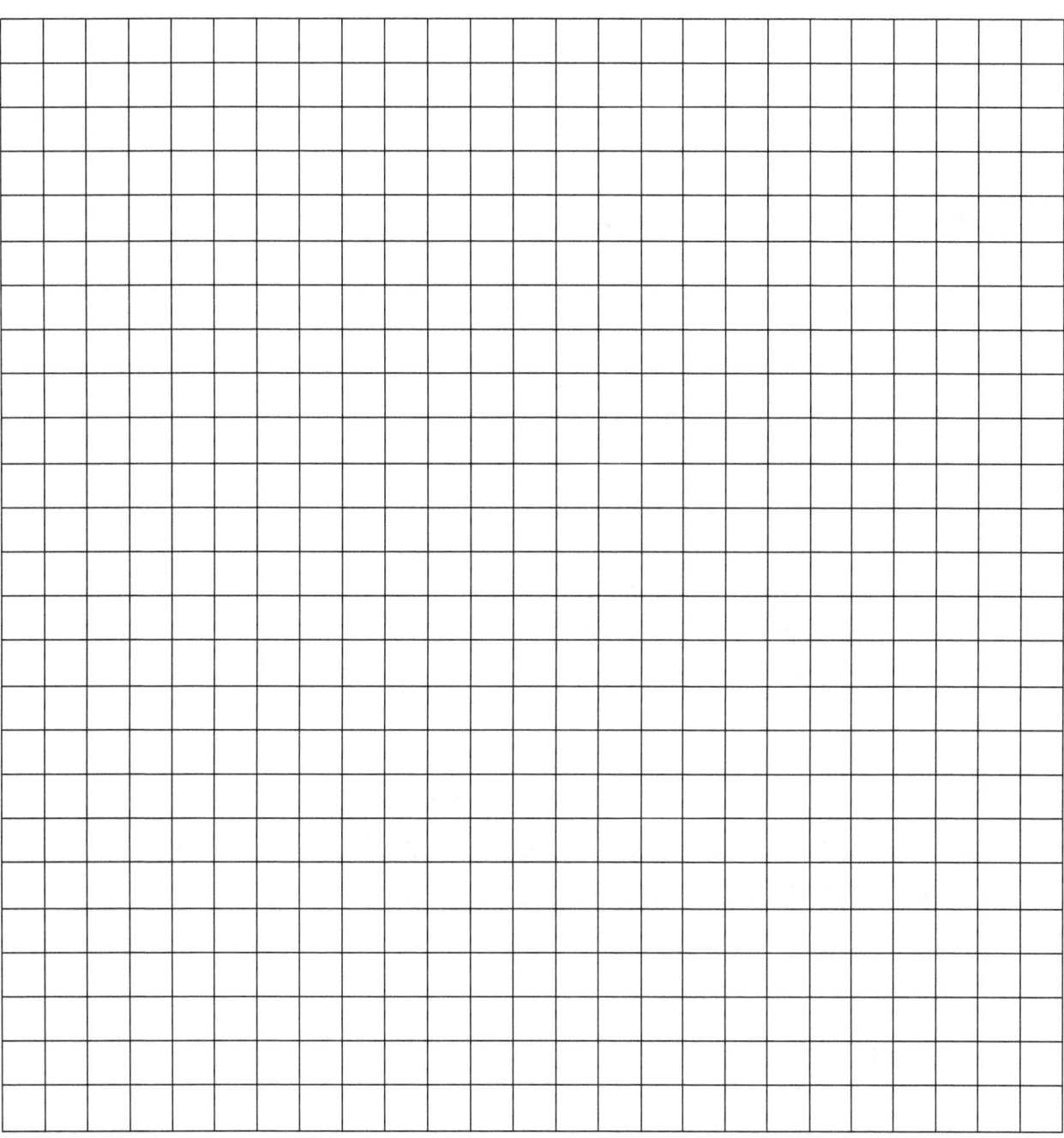

Questions:

1) What was the shape of each of those graphs?

2) Were all the measurements the same for all the leaves, seeds, and pinkies?

3) Why do you think populations have variation?

4) How can this be an advantage?

5) In natural selection, a certain trait is selected by nature to fit best in the environment. How can having a variety help a population survive when there is a sudden change in the environment?

6) Will the population look the same 1000 years after the environment's change selected out new successful individuals? (Explain)

7) If the population changes even slightly, evolution takes place. Does evolution take place if an individual dies? (Explain)

8) Does evolution take place if an individual is born? (Explain)

9) Where does evolution happen (to the individual or a population)? Explain.

10) How could big leaves become an advantage?

11) How could big leaves become a disadvantage?

12) How could long toes become an advantage?

13) How could long toes be a disadvantage?

14) How does variation set up a population for speciation?

15) What could be some sources of error in this investigation?

Changing Environments for Beads

Directions:

You will need **red**, **white**, and **blue beads** in a **bowl, red, white,** and **blue construction paper**, and **colored pencils. Looking at the materials and lab we will be using, what are the safety precautions we should take to protect ourselves and materials during the investigation?**

1) Have each group randomly get 10 beads out of the bowl and place them on white paper. The beads will represent a population with three variants in them, and the paper will represent the environment they are in. Count how many beads there are of each color and write this in Data Table 1 for the first generation.

2) The students will represent a predator of the beads. Have the students in each group take out three beads that do not match the environment's background, place them back into the bowl, and randomly pick three more beads out of the bowl (if you do not have any that don't match the background take ones that do match until you have three).

3) Add the three new beads to the paper, count how many beads there are for each color in the population, and write this down in Data Table 1 for Generation 2.

4) Repeat steps 2 and 3 for six more generations.

5) After completing eight generations change the white background to red or blue and predict how your population will change over time.

6) Repeat steps 2 and 3 for eight generations and write this data in Data Table 2.

7) Once you have completed both Data Tables, graph your data for Data Table 1 on Graph 1 and Data Table 2 on Graph 2, making a line graph using red (for red beads), black (for white beads), and blue (for blue beads) colored pencils.

8) Once the graphs are completed, answer the questions that follow.

Data Table 1

Bead Color	Gen 1	Gen 2	Gens 3	Gen 4	Gen 5	Gen 6	Gen 7	Gen 8
Red								
White								
Blue								

Data Table 2

Bead Color	Gen 1	Gen 2	Gen 3	Gen 4	Gen 5	Gen 6	Gen 7	Gen 8
Red								
White								
Blue								

Graph 1 (Line Graph) Color of Background: White

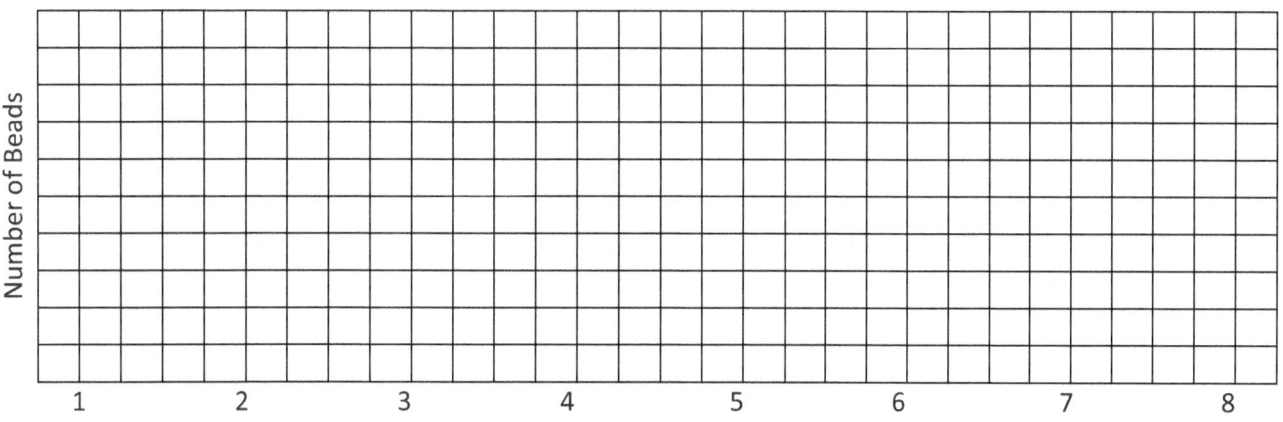

Number of Generations

Graph 2 (Line Graph) Color of Background: _____

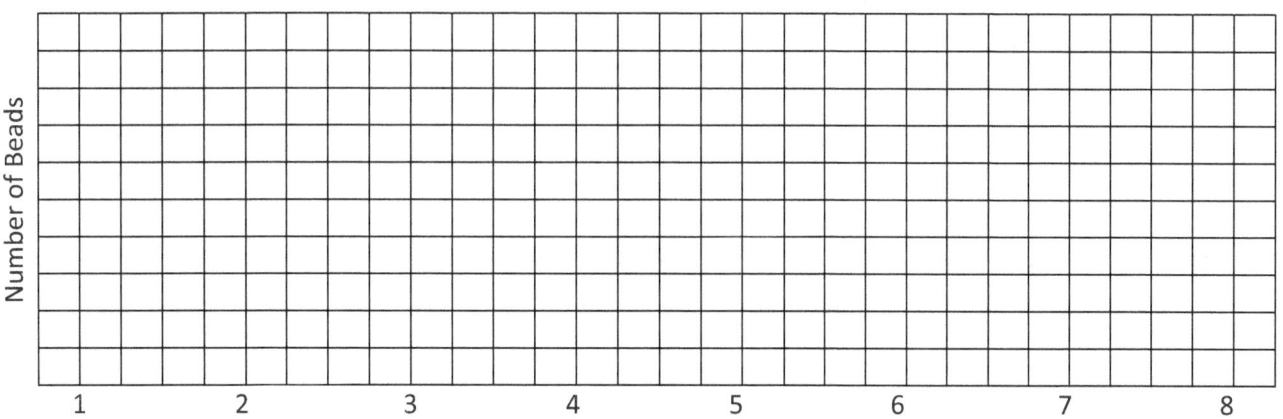

Number of Generations

Questions:

1) How did the population of beads change when there was a white background?

2) Why would this be useful in nature?

3) How did the population of beads change when there was a red or blue background?

4) Why would variations in a population benefit a population when an environment changes?

5) How could having variations in a population be hazardous to some individuals in the population?

6) What could happen to this population if the background turned green?

7) Which population would be more fit, one with little variation or one with lots of variation? Explain why.

8) How was this a good model for showing how variations within the population help populations adapt to changing environments?

9) How was this model not accurate?

Goldfish Evolution

Directions:

You will need **food serving gloves** for the teacher, a **large mixing bowl**, **paper plates**, **cheese-flavored Goldfish Crackers**, and **pretzel flavored Goldfish Crackers. Looking at the materials and lab we will be using, what are the safety precautions we should take to protect ourselves and materials during the investigation?**

In this activity, students will represent predators, a goldfish-eating shark, which selectively preys upon goldfish in small populations. This shark likes to eat two kinds of fish: **yellow fish (cheese-flavored)** and **brown fish (pretzel flavored)**. The yellow fish are easy for you to see, so they are easy to catch and eat. Brown fish travel more quickly and can evade capture more easily. Because of this, you eat only yellow fish, unless there are no yellow fish around, in which case you eat the brown fish. Fish are replaced with individuals randomly selected from an ocean (mixing bowl full of Goldfish crackers). Brown fish is determined by the presence of a dominant allele (B), and yellow fish by a recessive allele (b).

1) Send one student from your group with a paper plate to collect a <u>random</u> population of 10 fish (crackers) from the mixing bowl (ocean). Your teacher will place them on your plate for you.
2) In data table 1, for generation 1, record the number of yellow and brown fish in the population.
3) Choose three yellow fish from the population and eat them. If you do not have any yellow fish, fill in the missing number by eating the brown fish for a total of 3 fish eaten.
4) Send one student to the bowl (ocean) to get three more random fish and add them to your population.
5) In data table 1, for generation 2, record the number of yellow and brown fish.
6) Repeat steps 3-5 until you have data for all five generations.

Data Table 1

Generations	#of Gold Fish	# of Brown Fish	% of Gold Fish	% of Brown Fish
1				
2				
3				
4				
5				

7) Using the information from Data Table 1 on page 131 to plot your data on the graph below to show how your population changed over time. For each generation, plot two separated bars: use one color to represent the percent population of goldfish, and use a different color to plot the percent population of brown fish.

Graph 1

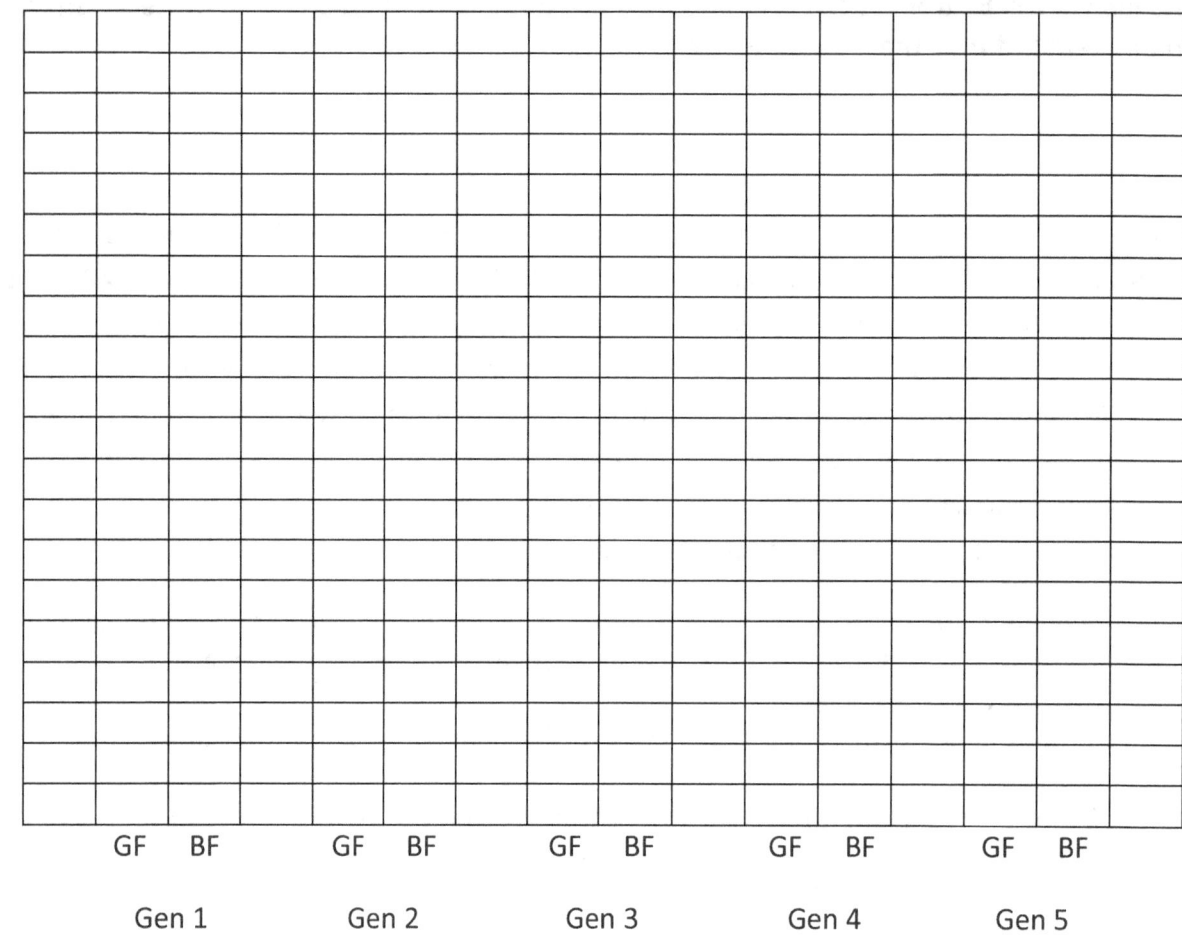

Questions:

1) How did the number of yellow fish change from generation 1 to 5?

2) Which phenotype was reduced in this population over time? Why?

3) What event occurs if there is a change in a population over time?

4) Explain what would happen over time if the brown fish were easier to catch?

5) What would happen if both fish were equally easy to catch?

6) How does this model show the Theory of Evolution and how speciation can occur?

Domains

Directions:

Use the **internet** and your **textbook** to research the Domain category in the Linnaean Classification system. Use that information to answer the following questions.

1) What are the two domains in life?

2) How are they separated?

3) What are the characteristics of each?

4) Which one came first?

5) How did the second one form?

6) Which domain do we belong to and why?

7) Which types of organisms belong to each domain?

8) Which domain has greater biodiversity? Explain.

9) Which domain is found in extreme environments? Give examples.

10) Which domain do you think would more likely be found on another planet? Explain why.

Functions the Kingdoms Serve

Directions:

Use your **textbook** and the **internet** to find the different functions organisms from these kingdoms serve in ecosystems and record their diversity (number of species in that kingdom). Then answer the questions that follow.

Kingdoms	Functions	# of Species
Archaea Bacteria		
Eubacteria		
Protista		
Fungi		
Plantae		
Animalia		

Questions:

1) Which kingdoms collect energy from the sun to start food webs (are producers)?

2) Which kingdoms produce oxygen for the atmosphere?

      a. Which kingdom provides the most oxygen?

3) Which kingdoms have species that support other organisms by living inside those organisms? Give examples.

4) Which kingdoms have species that recycle nutrients from dead organisms? Give examples.

5) Which Kingdoms have species that cycle nutrients?

6) Which kingdoms collect and use energy?

7) Which kingdoms eat (are heterotrophs)?

8) Which kingdoms can cause disease? Give some examples.

9) Which kingdoms are unicellular?

10) Which kingdoms are multicellular?

Relationships in Classification

Use the diagram below to see how organisms are classified from Kingdom to Species. Use this diagram and the **internet** to answer the questions that follow.

Diagram 1

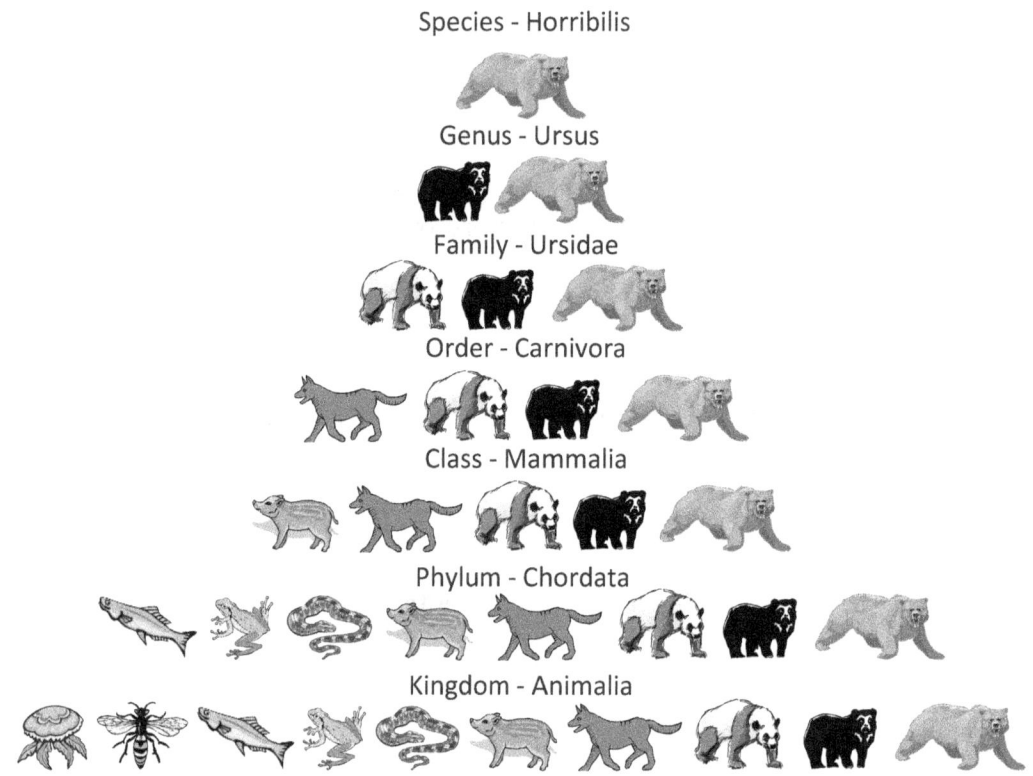

1) Tell the characteristics that define each group.

    a. Kingdom

    b. Phylum

    c. Class

    d. Order

    e. Family

    f. Genus

    g. Species

2) What happens to the diversity from the top of Diagram 1 to the bottom of Diagram 1?

3) In the Pyramids below, label each taxonomic level for the Bear seen in Diagram 1 and Humans. You may need to do a little research on the internet to fully classify ourselves as humans to see where we fit into the world.

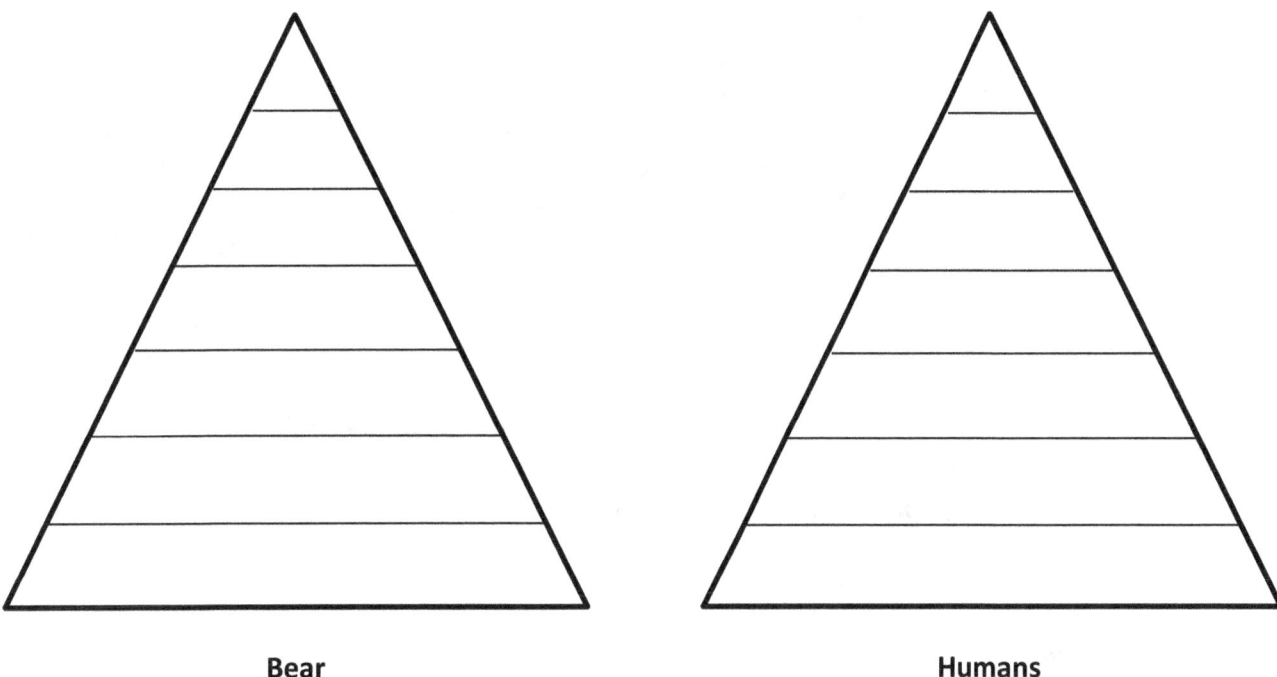

Bear   **Humans**

4) According to your human pyramid, are humans animals? Why?

5) Hominids are apes that stand upright. Are there any other hominids still around today that are not humans?

6) Were there any other Hominids in the past?

7) Which categories do we share with bears?

8) What characteristics separate us from bears?

9) Label the Venn diagram with these organisms by placing the letter of terms inside the appropriate circle (an example is already there):

a. Primates
b. Roaches
c. Invertebrates
d. Humans
e. Animals
f. Mammals
g. Insects
h. Vertebrates

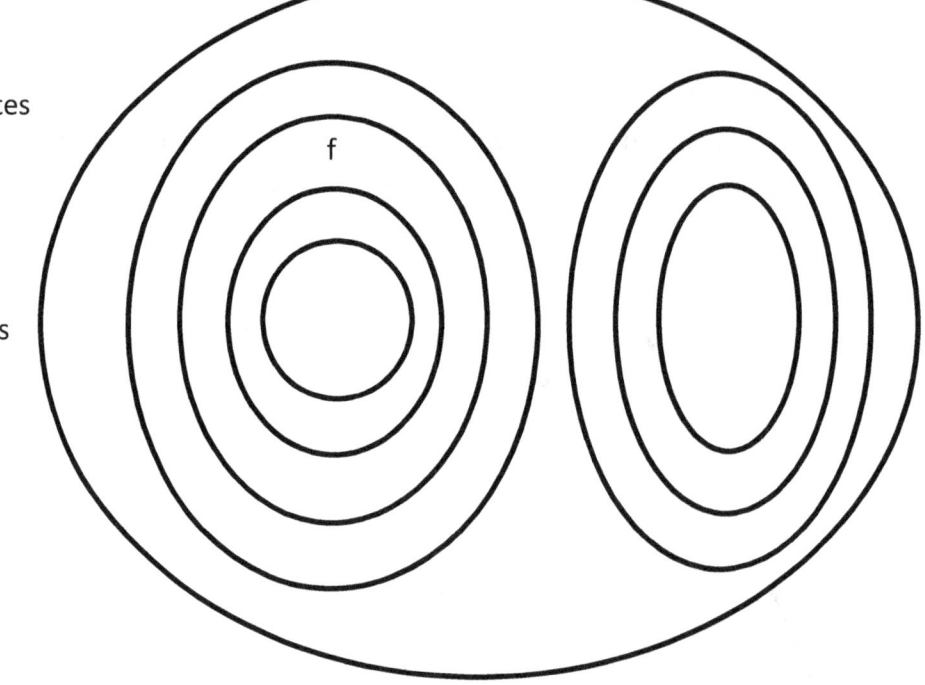

10) Which category contains both primates and roaches?

11) Are roaches vertebrates or invertebrates?

12) Which group in # 9 is the most diverse? (Explain why)

13) Which group in # 9 is the least diverse? (Explain why)

Pamishan Dichotomous Key

Help! Scientists have discovered quite a few new creatures on the planet Pamishan. They need your help to identify and classify them. Start with the first alien and use the dichotomous key on the next page. After identifying one alien, move on to the next alien until ALL have been identified. Put the scientific names using binomial nomenclature on the Pamishan Creatures Answers page at the end.

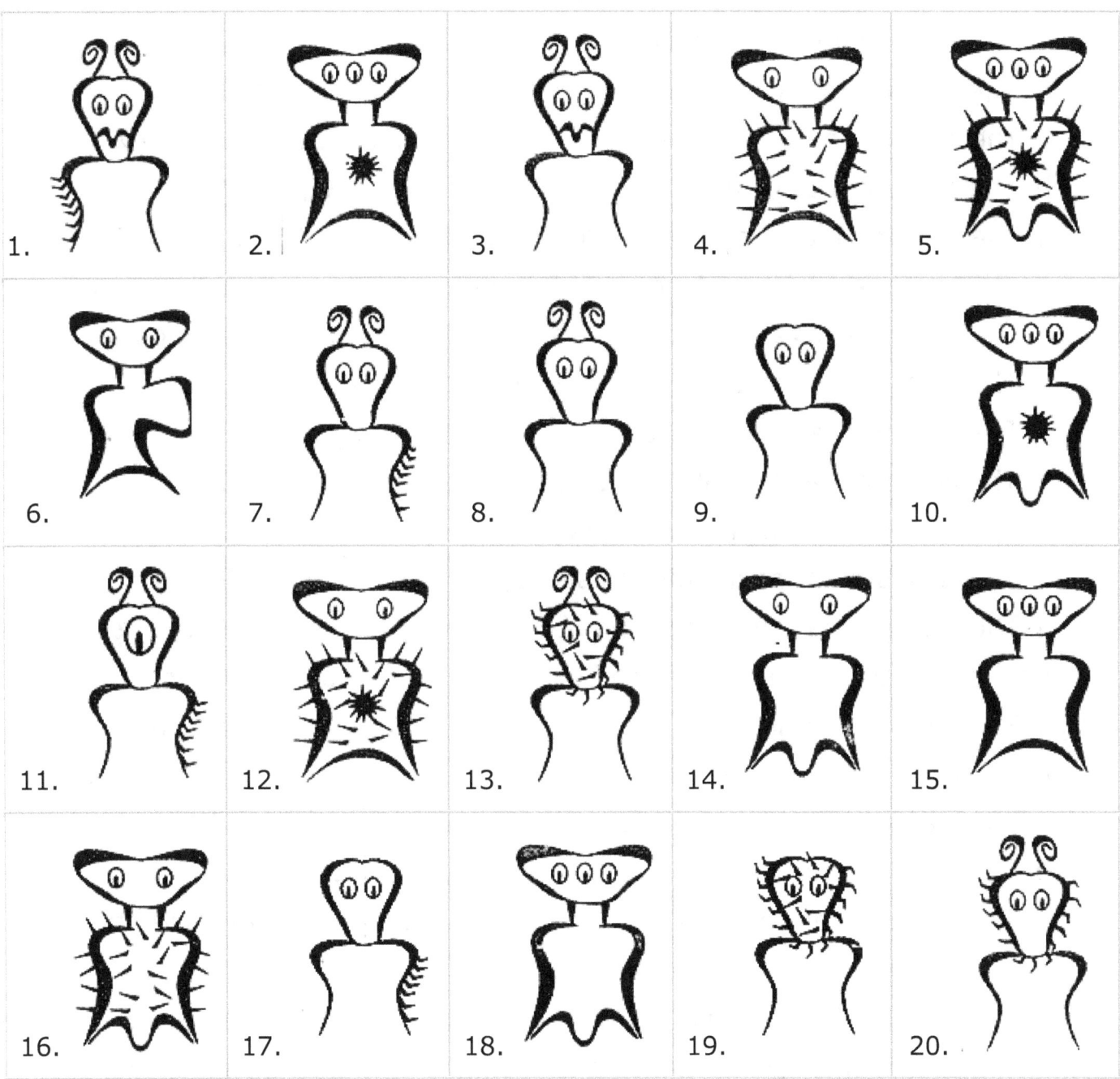

A Dichotomous Key to Pamishan Creatures

1. a. The creature has a large wide head............................go to 2

    b. The creature has a small narrow head.......................go to 11

2. a. It has 3 eyes ....go to 3

    b. It has 2 eyes ....go to 7

3. a. There is a star in the middle of its chest.....................go to 4

    b. There is no star in the middle of its chest ....go to 6

4. a. The creature has hair spikes ....*Broadus hairus*

    b. The creature has no hair spikes..................................go to 5

5. a. The bottom of the creature is arch-shaped...............*Broadus archus*

    b. The bottom of the creature is M-shaped...................*Broadus emmus*

6. a. The creature has an arch-shaped bottom..................*Broadus plainus*

    b. The creature has an M-shaped bottom......................*Broadus tritops*

7. a. The creature has hairy spikes ....go to 8

    b. The creature has no spikes..go to 10

8. a. There is a star in the middle of its body ....*Broadus hairystarus*

    b. The is no star in the middle of its body ....go to 9

9. a. The creature has an arch-shaped bottom ....*Broadus hairyemmus*

    b. The creature has an M shaped bottom ....*Broadus kiferus*

10. a. The body is symmetrical ....*Broadus walter*

    b. The body is not symmetrical.......................................*Broadus anderson*

11. a. The creature has no antennae ....go to 12

    b. The creature has antennae ....go to 14

12. a. There are spikes on the face ....*Narrowus wolfus*

    b. There are no spikes on the face ....go to 13

13. a. The creature has no spike anywhere ....*Narrowus blankus*

    b. There are spikes on the right leg ....*Narrowus starboardus*

14. a. The creature has 2 eyes...go to 15

b. The creature has 1 eye...*Narrowus cyclops*

15. a. The creature has a mouth.......................................go to 16

b. The creature has no mouth......................................go to 17

16. a. There are spikes on the left leg ....*Narrowus portus*

b. There are no spikes at all ....*Narrowus plainus*

17. a. The creature has spikes ....go to 18

b. The creature has no spikes ....*Narrowus georginia*

18. a. There are spikes on the head ....go to 19

b. There are spikes on the right leg.............................*Narrowus montanian*

19. a. There are spikes covering the face ....*Narrowus beardus*

b. There are spikes only on the outside edge of head..*Narrowus fuzzus*

Pamishan Creatures Answers

1) _____

2) _____

3) _____

4) _____

5) _____

6) _____

7) _____

8) _____

9) _____

10) _____

11) _____

12) _____

13) _____

14) _____

15) _____

16) _____

17) _____

18) _____

19) _____

20) _____

Classifying Animals

Directions:

1) Look at the pictures of the extinct animals and use them to fill in Data Table 1. Put an "**X**" in the box if the animal has that feature; leave it blank if it does not have that feature. Keep in mind **forelegs** are the front legs or legs closest to the head. Fish and reptiles here have **scales**; birds have both **scales** and **feathers**; mammals have **hair**. Check **smooth skin** if the animal does not have **scales**, **hair**, or **feathers**.

2) Then use the pictures of the extinct animals to fill in the dichotomous key. Keep in mind, Birds and mammals are **endothermic.** Which means they produce their own body heat or are considered to be warm-blooded. Fish, amphibians, and reptiles are **ectothermic.** They rely on the environment to give them their body heat or are considered to be cold-blooded.

Data Table 1

Animal	Fins	Wings	Forelegs	Hind legs	Horns	Smooth Skin	Scales	Feathers	Hair	Gills	Lungs
Domed Tortoise											
Dodo Bird											
Sculpin											
Texas Red Wolf											
Passenger Pigeon											
Elk											
Island Boa											
Bull Frog											
Bison											
Grayling											

Passenger Pigeon

Utah Lake Sculpin

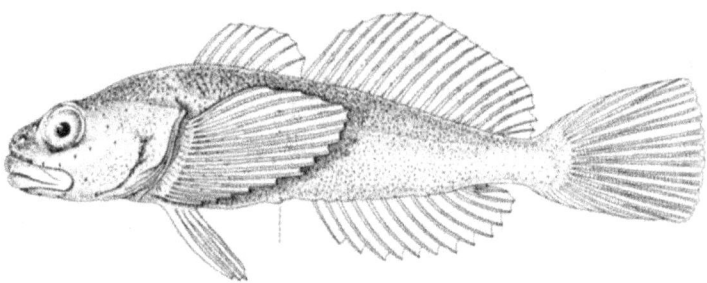

Figure 10. Utah Lake sculpin, *Cottus echinatus*. An adult female, 64.5 mm SL, collected from Utah Lake at the mouth of the Provo River, Utah County, Utah, during April 1928. *Drawing by Suzanne Runyan.*

Domed Tortoise

Elk

Bison

Texas Red Wolf

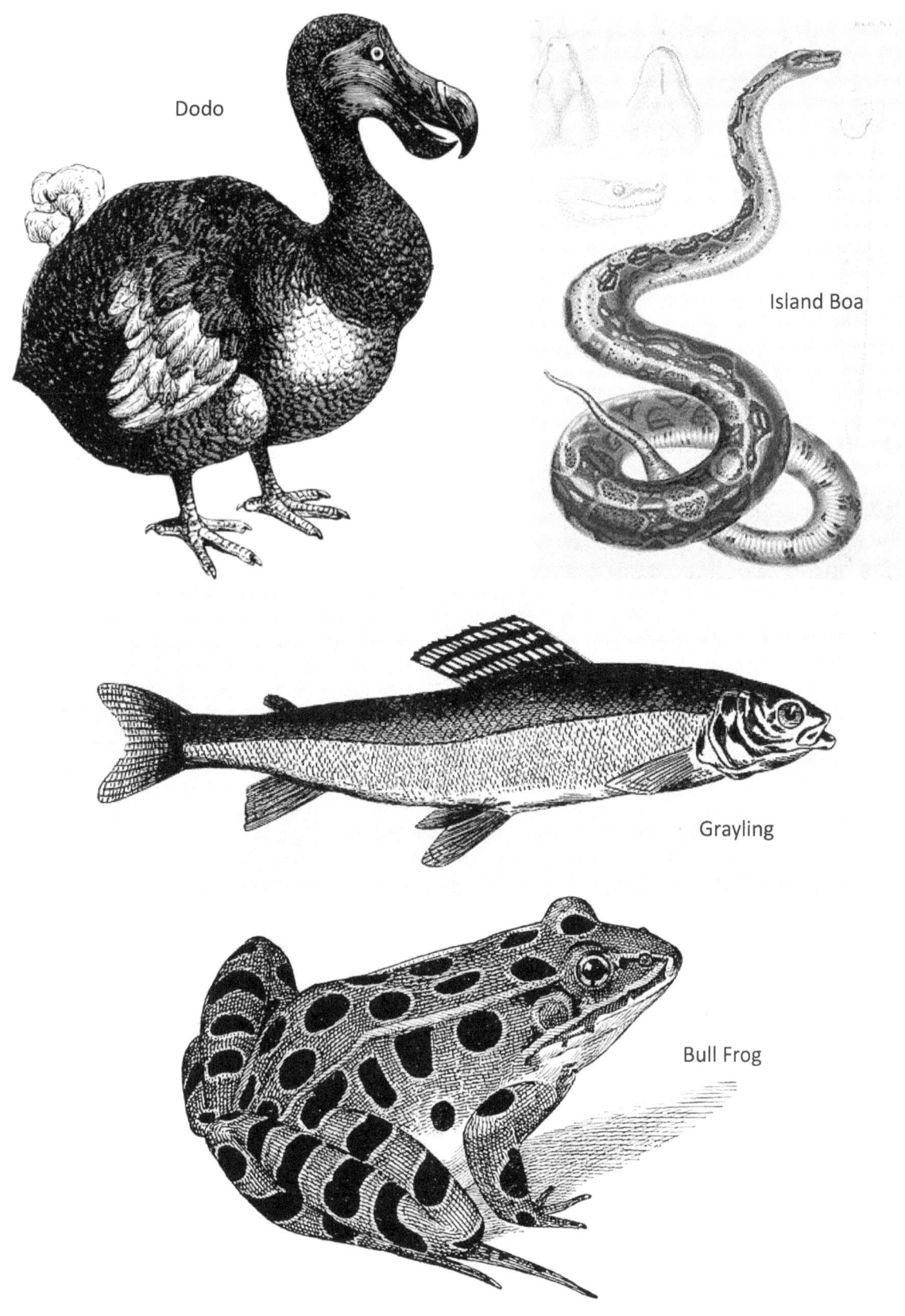

Dodo

Island Boa

Grayling

Bull Frog

Dichotomous Key for Animals

1a Is endothermic.. Go to 2

1b Is ectothermic... Go to 6

2a Has feathers... Go to 3

2b Has hair or fur.. Go to 4

3a Has narrow, straight beak...................................... _____

3b Has wide, crooked beak... _____

4a Has horns.. Go to 5

4b Has no horns... _____

5a Horns have many branches....................................... _____

5b Horns have no branches.. _____

6a Breathes with gills.. Go to 7

6b Breaths with Lungs.. Go to 8

7a Has large, fan-shaped fins just behind the head..... _____

7b Has small pectoral fins....................................... _____

8a Has scales covering skin....................................... Go to 9

8b Has smooth skin... _____

9a Has front and hind legs.. _____

9b Has no legs... _____

Questions:

1) Reptiles are ectothermic, have scaly skin, and breathe with lungs. Which of the animals are reptiles?

2) The Bull Frog is an amphibian. How are amphibians different from reptiles?

3) Mammals are endothermic, have hair or fur, breathe with lungs, and give birth to live young. Which of the animals are mammals?

4) Birds and mammals are endothermic vertebrates. Which animals are birds?

5) Which of the following pairs of vertebrate groups are the most similar (share the most characteristics)? Circle your choice.

      a) birds and fish   b) amphibians and reptiles

      c) amphibians and mammals   d) fish and mammals

Virtual Investigations that go with Evolution

ExploreLearning.com

Building Pangaea Gizmo

Nuclear Decay Gizmo

Half-life Gizmo

Natural Selection Gizmo

Evolution: Mutation and Selection Gizmo

Evolution: Natural and Artificial Selection Gizmo

Dichotomous Keys Gizmo

Microevolution Gizmo

Hardy-Weinberg Equilibrium Gizmo

Rainfall and Bird Beaks Gizmo

Evolution STEM Case Gizmo

Evolution Handbook Gizmo

PhET.colorado.edu

Natural Selection

Radioactive Dating Game

Notes:

Unit 4 Prokaryotic Life

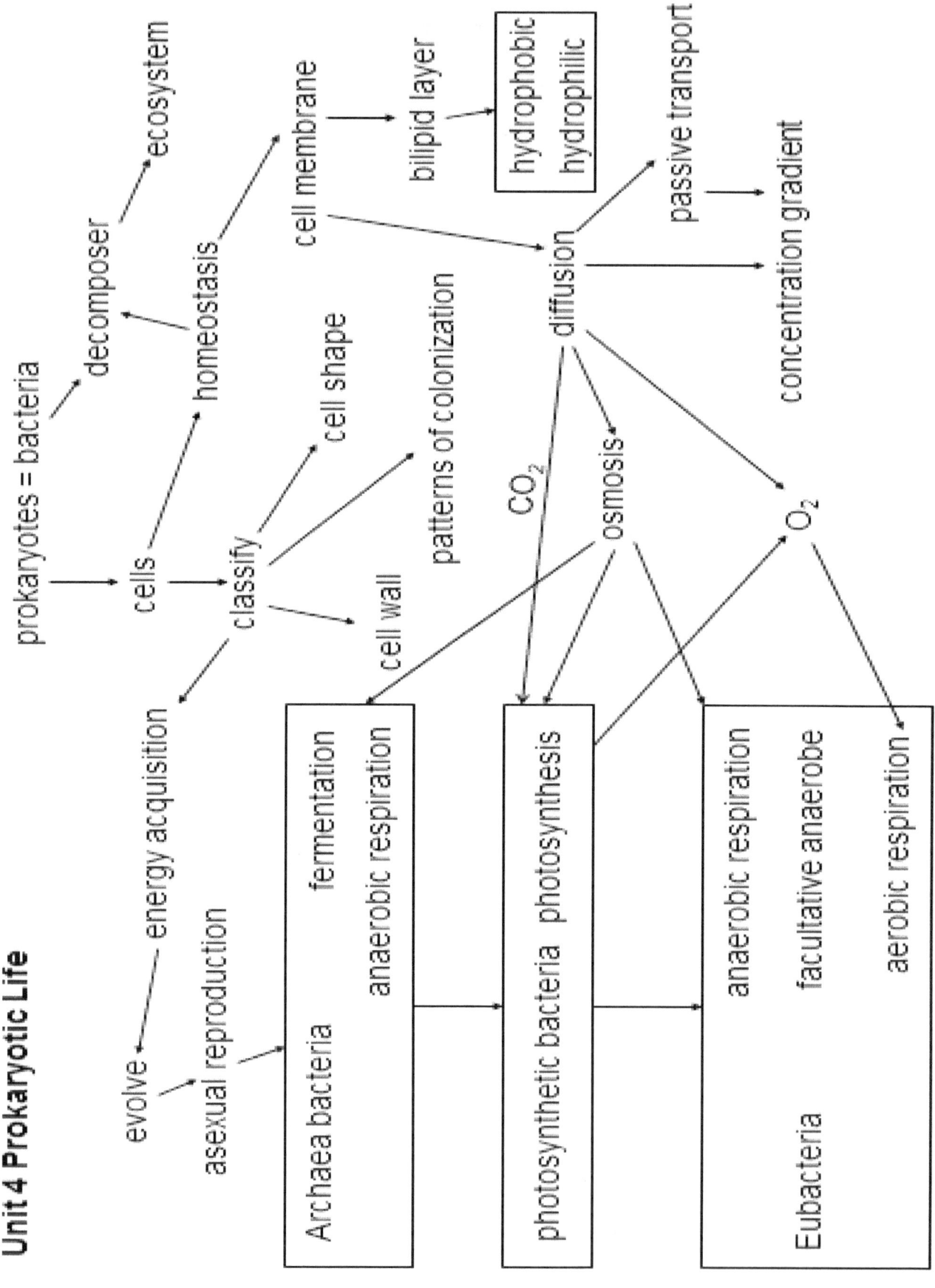

Unit 4 Prokaryotic Life

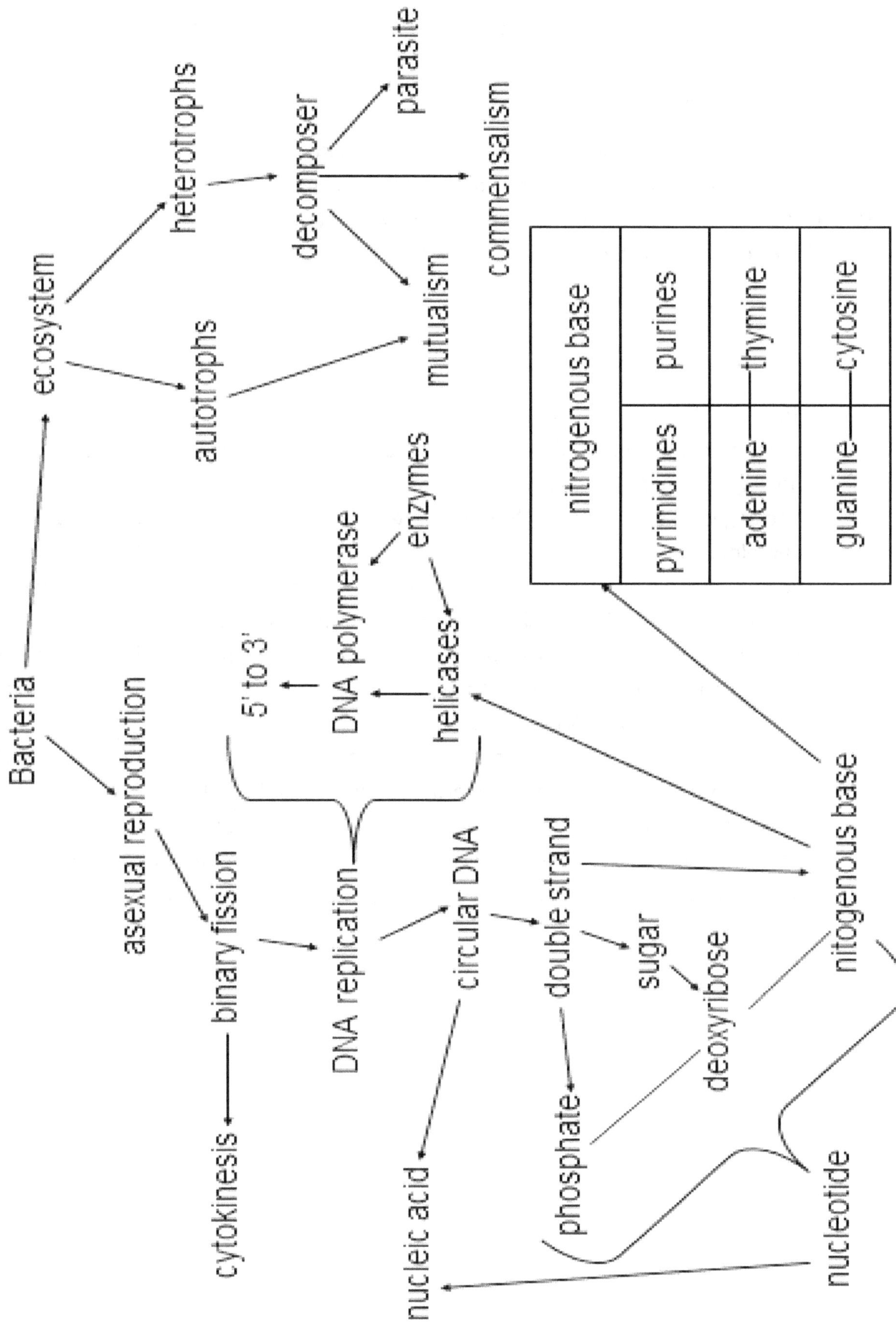

nitrogenous base	purines	adenine——thymine	guanine——cytosine
	pyrimidines		

Bacteria → ecosystem

ecosystem → heterotrophs
ecosystem → autotrophs

heterotrophs → decomposer

decomposer → parasite
decomposer → commensalism
decomposer → mutualism

autotrophs → mutualism

enzymes → DNA polymerase
enzymes → helicases

DNA polymerase → 5' to 3'

Bacteria → asexual reproduction

asexual reproduction → binary fission

binary fission → cytokinesis
binary fission → DNA replication

DNA replication → nucleic acid
DNA replication → circular DNA

circular DNA → double strand

double strand → sugar
double strand → nitogenous base

sugar → deoxyribose

phosphate
nucleotide → nitogenous base

Effects of Diffusion on Cells

Directions and Questions:

You will need a **sprig of** <u>**Anacharis elodea**</u>, two **eye droppers/pipets**, a **compound microscope**, a **slide** and **coverslip**, a **small beaker** of **fresh water**, a **small beaker** of **salt water**, and a **paper towel**. **Looking at the materials and lab we will be using, what are the safety precautions we should take to protect ourselves and materials during the investigation?**

1) Take one leaf from the elodea sprig and place it on the slide. Add a few drops of fresh water to the elodea on the slide. Hold the coverslip to not make fingerprints on it by holding the opposite edges between your thumb and finger. Hold the coverslip so that one edge hits the slide away from the water on the slide, then slide it toward the water and elodea until water hits and slides across the lowered edge of the coverslip. Then gently lower the other side of the coverslip down so that few bubbles are trapped under the coverslip. If this is confusing, have your teacher show you.

2) Place the slide on the microscope and focus it under low power. Follow your teacher's instructions on how to center it and which knobs to use when focusing. When in focus, it should look like a brick wall.

3) Then switch your lens to medium power and focus it according to your teacher's instructions. You should see the chloroplasts in each of the cells.

4) Then switch your lens to high power and focus it using your teacher's instructions. See the chloroplast possibly moving inside the cells. Draw a picture below of the cells you see under the microscope; this is an example of how plant cells look in a **hypotonic** solution.

5) Now take up some salt water into the other pipet and gently squirt a little on the slide just left of the coverslip. Place the paper towel's edge under the coverslip's right edge to pull out the freshwater allowing the saltwater to be pulled under the coverslip. Watch as you do this. What do you see happen to the cells?

6) Draw a picture of what the cells look like now. You should see the cell membrane pulled away from the cell wall, making it look like little balls inside each cell. This image shows how a cell behaves in a **hypertonic** solution.

7) Did the cells get bigger or smaller?

8) What kept them from taking in more water and exploding like a water balloon when the cells were in fresh water?

9) Animal cells do not have this structure, so the animal's body must regulate how much water can be in the body, making it **isotonic**. What would happen if too much water was allowed in a solution (**hypotonic**)?

10) What would happen to the cells if there was too little water in the animal (**hypertonic**)?

Osmosis Toothpicks

Directions:

You will need five **toothpicks**, a **pipette**, and **water. Looking at the materials and lab we will be using, what are the safety precautions we should take to protect ourselves and materials during the investigation?**

1) Take your five toothpicks and bend them in half, not separating them. Place them on the table where the bent areas face inward towards each other in a circle forming a star.
2) Fill your pipette full of water, squirt it on the elbows of the bent toothpicks, and observe what happens.

Questions:

1) What did you see when you added the water?

2) Why do you think this happened?

3) Wood is made up of xylem tissue in trees that carries water to the rest of the plant. How did the toothpicks show the xylem was transporting water?

4) Osmosis is the diffusion of water. How did you see osmosis take place in this investigation?

Membrane Models

Directions and Questions:

You will need a **large beaker of water**, **food coloring**, **dishwashing liquid**, **long twisty ties**, a **small shallow dish**, **toothpicks**, **strainer** or **colander**, a **small beaker of water**, **salt**, **marbles**, **dry beans**, a **small tub** or **bucket**, and two **pieces of paper**. **Looking at the materials and lab we will be using, what are the safety precautions we should take to protect ourselves and materials during the investigation?**

Part 1 Diffusion and Osmosis

1) Take the large beaker of water and drop a couple of drops of food coloring in the water. How do you see the food coloring move?

2) When did it seem to stop moving?

3) You just watched the process of **diffusion**. And the diffusion of water is **osmosis**. These are examples of **passive transport**. The next two sections of the lab will show you how some materials can passively move across a cell membrane and others cannot.

Part 2 Semipermeable by Material

4) Cover the bottom of the small dish with soap. Take two twisty ties. Make a large loop with the first tie; make a much smaller loop with the second tie. Bend the rest of each tie up at a 90° angle to use as a handle to dip into and pull out of the soap.

5) Now dip the loops into the soap and pull it out. Notice it produces a thin soapy membrane that stretches across the loop. Which loop holds the membrane on it longer? This model should show us why cells cannot grow bigger; the membrane cannot hold together a large cell.

6) Now dip the small loop into the soap and pull it out with a soapy membrane. Take a toothpick and poke it through the membrane. What happens to the membrane?

7) This time repeat the procedure in #6 but this time, wet the toothpick with water, then gently poke it through the soapy membrane. What happened to the membrane this time?

8) This time repeat the procedure in #6 but take a new toothpick, cover one end of it with soap, and then gently poke it through the soapy membrane. What happened to the membrane?

9) Make sure the small loop is lined with soap. Dip the large loop into the soap and pull it out with a soapy membrane. Now try to pass the small loop through the larger loop with the membrane while holding it parallel to the ground. What do you notice? This process makes an opening act as a protein channel, allowing materials to move in and out of the cell membrane.

10) What was used as a cell membrane model in this part of the investigation?

Part 3 Semipermeable Membrane by Size

11) Observe the colander; what does it have all over it that allows some things to pass through and other things not pass through? These represent the protein channels in the cell membrane.

12) Take the water, salt, marbles, and dry beans and see which substances can pass through as you pour them in over a small tub or bucket. Which materials were allowed to pass through?

13) Which materials were not allowed to pass through?

14) Why did those materials not pass through? This model shows why many molecules do not pass through in and out of cell membranes; this is why cell membranes are semipermeable, allowing some things to pass through and stopping others.

15) What was used as a cell membrane model in this part of the investigation?

Part 4 Active Transport

16) When molecules are too big to pass through, many substances can be recognized by the cell membrane and act as your classroom door. A flat piece of paper can slide under the door. Slide the flat paper under the door. Did it go under the door?

17) A wadded paper cannot pass under the door. Wad the second paper and try to have it pass under the door. Was the wadded paper able to pass under the door?

18) But a large protein (like me) can trigger the doorknob to open it up to allow me to pass through. There are structures outside the cell membrane that trigger **endocytosis** allowing large substances to enter the cell.
   a. When it eats, it is called **phagocytosis,** forming a food vacuole.
   b. When it drinks large amounts of water all at once, this is called **pinocytosis,** forming a water vacuole.

     c.   What type of endocytosis is this modeling when you pass through the door?

19) How do you think a large protein made inside the cell is allowed to pass out of the cell during **exocytosis** (hint: remember the soap)?

20) Was energy needed to open the door?

21) What was used as a cell membrane model in this part of the investigation?

22) You can think of **passive transport** as moving with an escalator. Do you have to use any energy to move with the escalator?

23) You can think of **active transport** as trying to move against an escalator. Do you have to use energy to move against an escalator?

24) Describe how cell membranes are used in active and passive transport.

Seeing Different Types of Bacteria

Directions and Questions:

You will need **prepared slides of round, rod, spiral-shaped bacteria**, and **protists** like an **amoeba**, **euglena**, or **paramecium**. You will use a **compound light microscope** to see them. You may also want a **lens wipe** to clean off the slides, lenses, and eyepiece. **Looking at the materials and lab we will be using, what are the safety precautions we should take to protect ourselves and materials during the investigation?**

1) Place the coccus bacteria slide (round-shaped bacteria) on the microscope and focus it under low power. Follow your teacher's instructions on how to center it and which knobs to use when focusing.
2) Then switch your lens to medium power and focus it according to your teacher's instructions. You might see tiny dots of bacteria.
3) Then switch your lens to high power and focus it using your teacher's instructions. Draw a picture of the cells that you see under the microscope.

4) Repeat the procedure for #s 1-3 for the bacilli bacteria (rod-shaped bacteria). Draw a picture of the Bacillus bacteria.

5) Repeat procedure #s 1-3 for Spirillum (spiral-shaped bacteria). Draw a picture of the Spirillum bacteria.

6) Now compare the size of a single-celled eukaryote to the bacteria. Repeat the procedure of #s 1-3 for a protist. Draw a picture of an amoeba, euglena, or a paramecium.

7) Estimate how many bacteria could fit inside this eukaryotic cell.

8) Did you notice any structures inside the eukaryotic cells? Describe how they looked.

9) Many of the organelles inside Eukaryotes came from or are bacteria themselves, like mitochondria and chloroplasts. Knowing this, how do you think eukaryotes could have evolved from prokaryotes?

Seeing Live Bacteria in Yogurt

Directions:

You will need a **compound microscope**, **plain yogurt**, a **toothpick**, a **slide** and **coverslip**, a **pipet** or **eyedropper**, and **fresh water** in a **small beaker**. **Looking at the materials and lab we will be using, what are the safety precautions we should take to protect ourselves and materials during the investigation?**

1) Take your toothpick, dip it into the yogurt, and smear it on the slide. Place a drop or two of water on the yogurt on the slide. Then grab the coverslip so as not to make fingerprints on it by holding the opposite edges between your thumb and finger. Hold the coverslip so that one edge hits the slide away from the water on the slide, then slide it toward the water and yogurt until water hits and slides across the lowered edge of the coverslip. Then gently lower the other side of the coverslip down so few bubbles are trapped under the coverslip. If this is confusing, have your teacher show you.

2) Place the slide on the microscope and focus it under low power. Follow your teacher's instructions on how to center it and which knobs to use when focusing.

3) Then switch your lens to medium power and focus it according to your teacher's instructions. You should see tiny dots. Make sure any dots are in the center of your field of view.

4) Then switch your lens to high power and focus it using your teacher's instructions. See the bacteria possibly moving (make sure the yogurt is warm at room temperature before starting the lab to see them move faster). Draw a picture of the cells that you see under the microscope.

Questions:

1) Describe how the bacteria seem to move.

2) What color are the bacteria?

3) What shape are the bacteria?

4) Are the bacteria arranged in any patterns? If so, describe what they are.

5) What evidence did you see that the bacteria in the yogurt are alive?

6) How did the bacteria in the yogurt get food?

7) How did the milk get turned into yogurt?

8) An average bacterium can reproduce every 20 minutes; this means it doubles its population every 20 minutes. If you leave food out for 4 hours, estimate how many bacteria will be on that food if only one bacterium were to fall on it initially.

DNA Replication

Directions:

You will need **red, black, green, and blue colored pencils, scotch tape, scissors,** and the **internet. Looking at the materials and lab we will be using, what are the safety precautions we should take to protect ourselves and materials during the investigation?**

1) Look at page 167. The dots represent the **phosphates;** the pentagon represents the **sugar (deoxyribose)**; these are the DNA molecule's sides; we will not color these. We will, however, color the **Adenine, Thymine, Cytosine,** and **Guanine.** Make sure you stay within the lines and just color the shapes of the nitrogenous bases. **Adenine** will be **red, Thymine** will be **black, Cytosine** will be **blue,** and **Guanine** will be **green.**

2) Once all the bases are colored, cut out all the pieces along the dashed lines. Leave the DNA strand together for now.

3) Now you are ready to act out the process of **DNA Replication**. Your scissors will act as the enzyme **DNA Helicase** as you cut down the dotted line and split the DNA molecule in half.

4) You need to take a strip of scotch tape and place half of it under one side of the DNA molecule where you just cut so that the new pieces will stick to the rest of the tape exposed as you bring the complementary bases in to build 2 DNA models. You will start on the 5' end and add the complementary bases (Adenine combines with Thymine, and Cytosine combines with Guanine) until you reach the 3' end. Then do the same for the other side of the DNA molecule you originally split. You are acting as the enzyme **DNA polymerase** as you do this.

5) Proofread your DNA strands to make sure you adhere to the **A-T/C-G rule**; this is also done by **DNA polymerase**. You can find someone else to proofread, if you want, to make sure your code is correct (they are just another **DNA polymerase** molecule helping you out).

6) Any uncorrected mistakes in the code are **point mutations** known as **substitutions**. At the end, the teacher can check for any mutations that made it through proofreading.

Questions:

1) When you are done, how many DNA molecules do you have?

2) How many DNA molecules did you have at the beginning?

3) Why do you think cells need to make a copy of their DNA before they reproduce and split in half? This reason is why DNA Replication is important. New cells would die without it.

4) Are the two sides of the DNA molecule facing the same or the opposite directions?

5) Look on the internet and see which bases are purines and which are pyrimidines. What is the shape of purines?

6) Which two bases are purines?

7) What is the shape of pyrimidines?

8) Which two bases are pyrimidines?

9) What is the purpose of DNA?

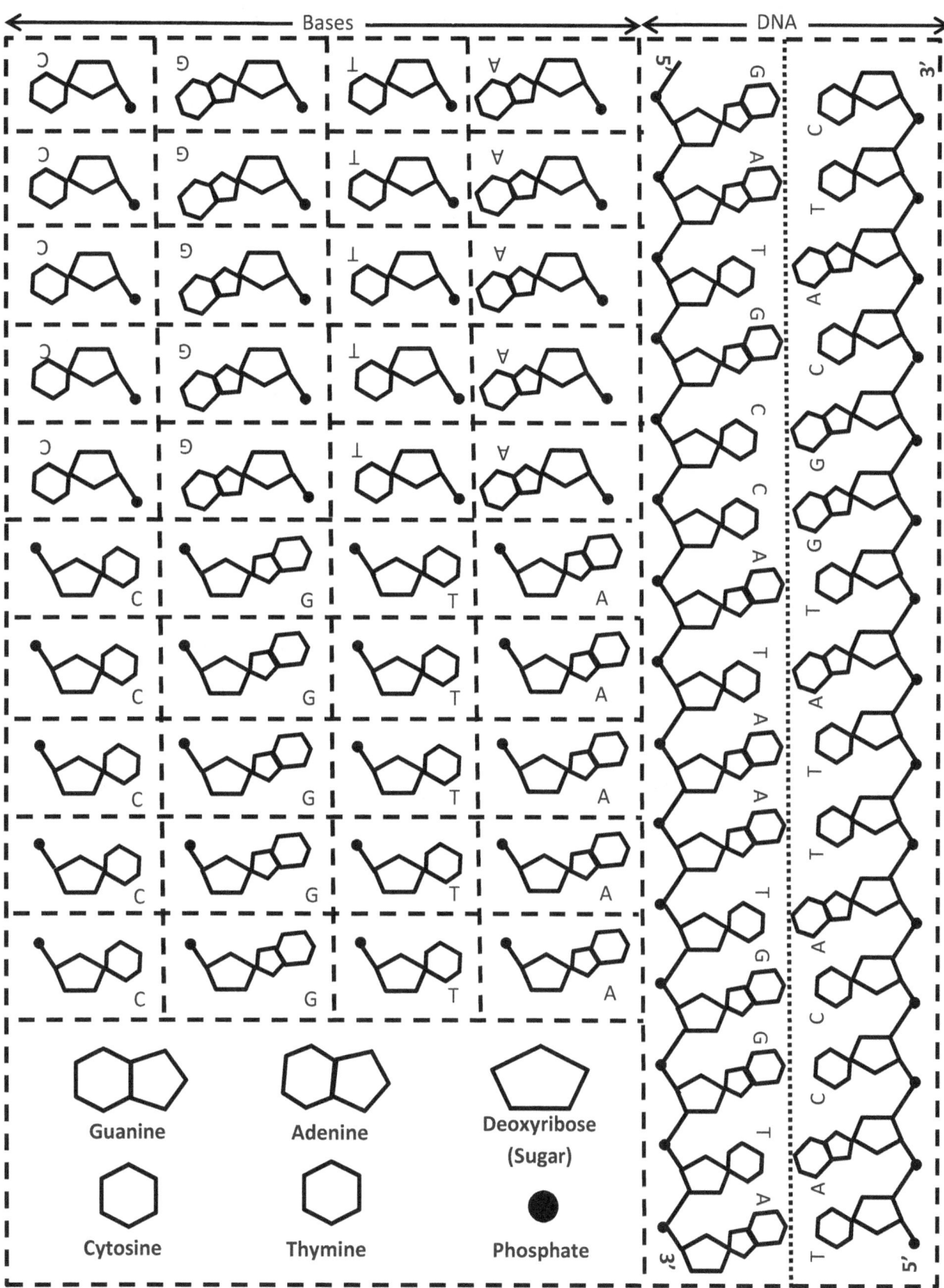

Bases

DNA

Guanine

Adenine

Deoxyribose (Sugar)

Cytosine

Thymine

Phosphate

This page will be cut up from the page before!

Draw a Detailed Picture of Bacteria

Directions:

1) Find a detailed picture of a bacterium on the **internet** or in your **textbook**, draw it, and label it below.

2) Tell the function of each part you labeled on the Bacteria.

Virtual Investigations that go with Prokaryotic Life

ExploreLearning.com

Diffusion Gizmo

Osmosis Gizmo

Building DNA Gizmo

Diffusion STEM Case Gizmo

Diffusion Handbook Gizmo

Osmosis STEM Case Gizmo

Osmosis Handbook Gizmo

PhET.colorado.edu

Membrane Channels

Unit 5 Unicellular Eukaryotes

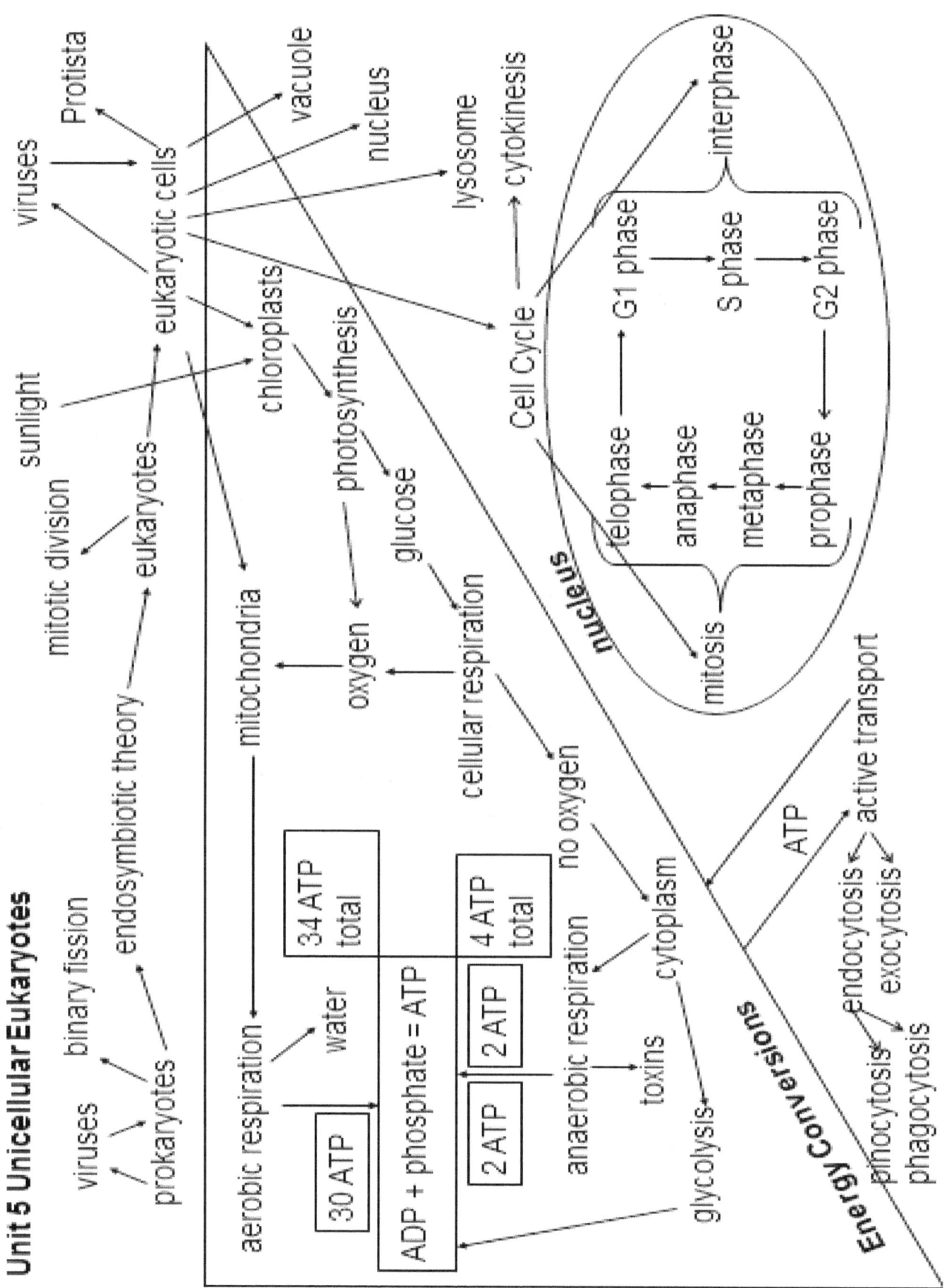

Unit 5 Compare and Contrast

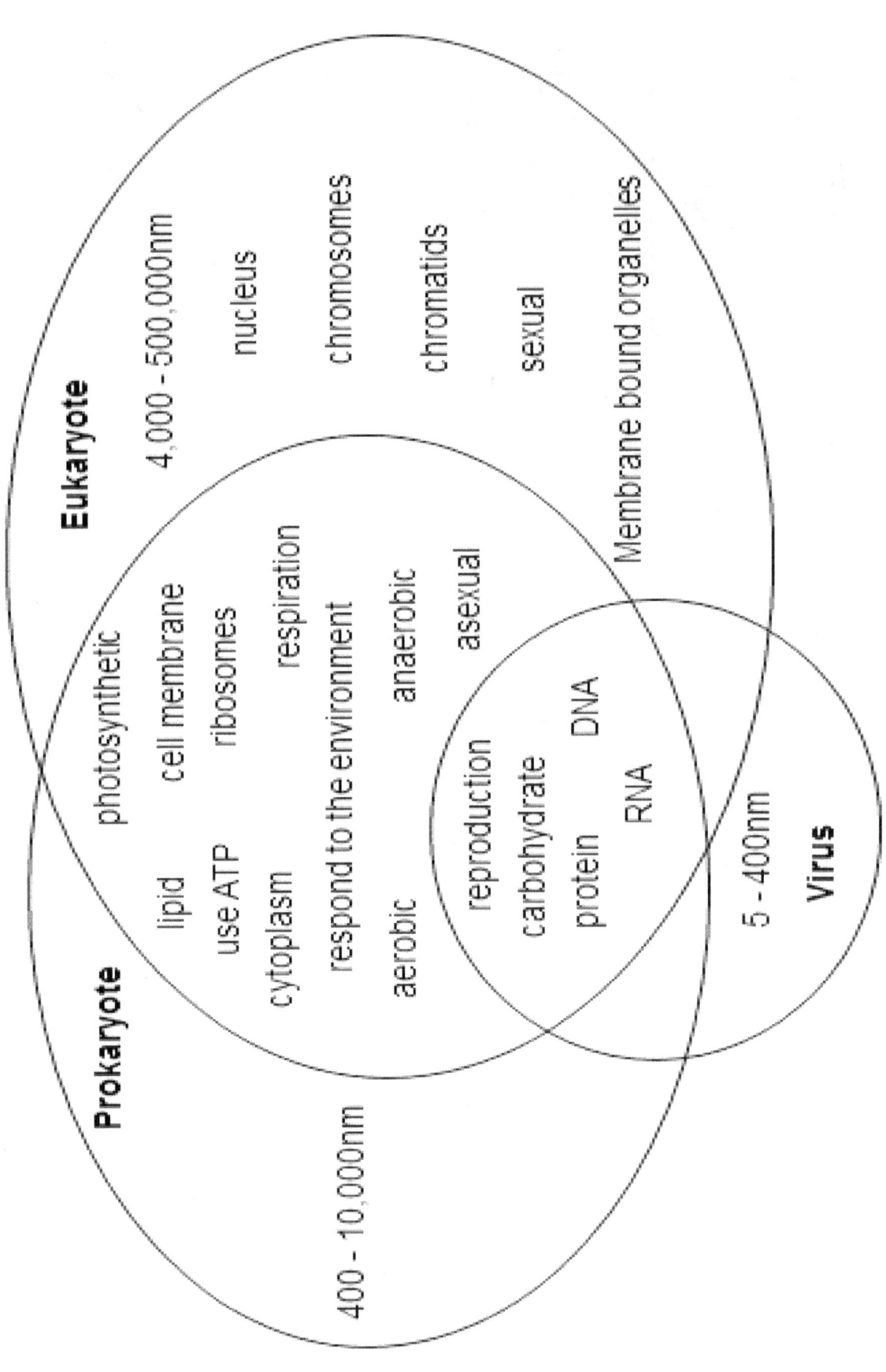

Eukaryote

4,000 - 500,000nm

nucleus

chromosomes

chromatids

sexual

Membrane bound organelles

Prokaryote

photosynthetic

cell membrane

lipid   ribosomes

use ATP   respiration

cytoplasm

respond to the environment

aerobic   anaerobic

asexual

reproduction   DNA

carbohydrate

protein   RNA

5 - 400nm

Virus

400 - 10,000nm

Cell Town

Directions:

Cell structures and functions are a lot like how a town is structured and functions. You need to draw a town and label each part of that town with a part of the cell that functions the same as that part of a town. This project can be done by hand or digitally. First, use the **internet** or your **textbook** to find each organelle's functions. Then find out what is in a town that would have the same functions. Write them down next to the organelles below. Then draw/build and label your town.

*For example, the **nucleus** is the cell's control center, so that would be like **City Hall**, so you would draw a **City Hall** and label it the **nucleus**.*

Cell wall

Cell Membrane

Nucleus – *City Hall*

Nucleolus

Nuclear Envelope

Chromatin/DNA

Ribosomes

Smooth Endoplasmic Reticulum

Rough Endoplasmic Reticulum

Golgi Apparatus

Lysosome

Chloroplast

Mitochondria

Cytoplasm

Cytoskeleton

Protein Channels

Water Vacuole

Food Vacuole

Chromoplast

Leucoplast

Characteristics of Prokaryotic and Eukaryotic Cells

Directions:

You will need **prepared slides of bacteria, Amoeba, Paramecium,** and **Euglena,** or you can prepare wet-mount slides of these same organisms according to your teacher's instructions. You will also need the **internet**, your **textbook**, a **compound light microscope, lens wipes**, and have your teacher pull up videos on **YouTube** of these organisms interacting and feeding for you to see. **Looking at the materials and lab we will be using, what are the safety precautions we should take to protect ourselves and materials during the investigation?**

1) Focus each of these organisms under your microscope and draw a picture of each in the table below. Show your teacher you can safely and correctly focus these organisms by having your teacher come by and look at one of them centered and focused on your microscope. Teacher's initials:

Table 1

Organism	Picture	Describe movement & feeding
Bacteria (Prokaryote)		
Amoeba (Eukaryote)		

Paramecium (Eukaryote)		
Euglena (Eukaryote)		

2) Look in your textbook or on the internet and label each organism's important parts on the pictures you have drawn in Table 1.

3) Search and watch YouTube videos on how each organism moves and feeds and describe them in Table 1.

Questions:

1) Based on your observations, do the cells have the same shape? Explain.

2) Based on your observations, do the cells have the same size? Explain.

3) Based on your observations, do all cells have the same parts? Explain.

4) What cell structures do you see are common to all cells?

5) What cell structures are found only in eukaryotic cells?

6) Why do you think different cells have different shapes and sizes?

7) How do the eyespot and chloroplast work together to help the euglena survive?

8) What characteristics of an animal does a euglena possess?

9) What characteristic of a plant does a euglena possess?

10) How is the amoeba like a white blood cell that engulfs invading organisms in our bodies?

11) How are the pseudopods used in amoeba?

12) What are the functions of the cilia on the paramecium?

13) Which of these organisms seem to be the most advanced/complicated? Explain.

14) Which of these organisms would probably be the easiest to keep alive in class if you had a solar light source?

15) How big do you expect the chloroplasts and mitochondria to be if they are bacteria that act as organelles in eukaryotes?

16) From what you observed in this lab, how does this give evidence for endosymbiosis?

17) Should we call this concept an Endosymbiotic Hypothesis or Theory? Explain.

Can You Make the Connection?

Directions:

You will need **plain** and **peanut M&Ms,** the **internet,** and a **textbook. Looking at the materials and lab we will be using, what are the safety precautions we can take to protect ourselves and materials during the investigation?**

1) We will model how plain M&Ms are like prokaryotic cells, and peanut M&Ms are like eukaryotic cells.
2) Describe the characteristics of prokaryotic cells in Data Table 1.
3) Describe the features of a plain M&M in Data Table 1.
4) Describe the characteristics of eukaryotic cells in Data Table 2.
5) Describe the features of a peanut M&M in Data Table 2.

Data Table 1

Object	Prokaryotic Cell	Plain M&M
Outside covering		
What is inside		
Comparable Size		
Components:		

Data Table 2

Object	Eukaryotic Cell	Peanut M&M
Outside covering		
What is inside		
Comparable Size		
Components:		

Questions:

1) How do you think plain M&Ms are like prokaryotic cells?

2) Explain how plain M&Ms are different from prokaryotic cells.

3) How do you think peanut M&Ms are like eukaryotic cells?

4) Explain how peanut M&Ms are different from eukaryotic cells.

5) Explain how prokaryotic cells are similar to eukaryotic cells.

6) Explain how prokaryotic cells are different from eukaryotic cells.

7) Use these terms to fill in the Ven Diagram comparing and contrasting prokaryotic and eukaryotic cells: **cell membrane**, **cytoplasm**, **DNA**, **cell wall**, **nucleus**, **movement**, **membrane-bound organelles**, and **ribosomes**.

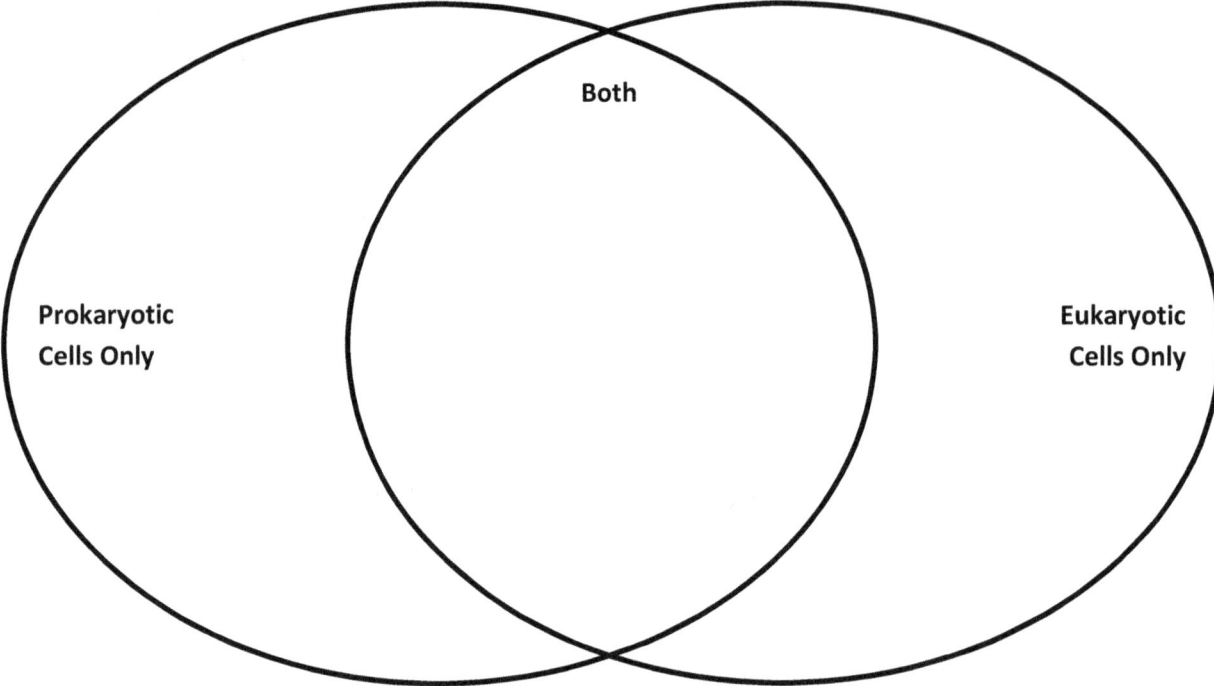

8) Use the internet or textbook to draw a scale model of a virus, a prokaryotic cell (bacteria), and a eukaryotic cell.

Seeing Cell Division

Directions:

To study mitosis stages, you will need pictures from your **textbook** to reference each stage of cell division, **prepared slides** of **Onion Root Tip** and **Whitefish Blastula**, a **compound light microscope**, and **lens wipes. Looking at the materials and lab we will be using, what are the safety precautions we should take to protect ourselves and materials during the investigation?**

1) For both the Onion Root Tip and Whitefish Blastula, find and draw a picture of each cell cycle stage in the following tables.
2) Then describe what important changes occur at each cell cycle stage in Tables 1 and 2.

Table 1: Onion Root Tip

Stages	Picture	Description of Changes
Interphase		
Early Prophase		
Late Prophase		

Metaphase		
Anaphase		
Early Telophase		
Late Telophase		
Cytokinesis		

Table 2: Whitefish Blastula

Stages	Pictures	Description of Changes
Interphase		

Early Prophase		
Late Prophase		
Metaphase		
Anaphase		
Early Telophase		
Late Telophase		
Cytokinesis		

Questions:

1) Compare what happens during prophase and telophase; how are they similar to each other?

2) Why does Mitosis need to occur for cells to survive after cell division?

3) What evidence did you see that shows mitosis is a continuous process, not a series of separated events?

4) During which stage of the cell cycle does DNA Replication occur (also known as DNA Synthesis)?

5) Why does DNA Synthesis need to happen?

6) Why do organisms need cell division to take place?

7) What would happen if the process of mitosis skipped metaphase and anaphase?

8) What happens if cell division gets out of control and happens too fast?

9) What do we call this?

Hand Models Showing Cell Division

Directions:

Follow the instructions for this model with your hands. You can do this together with your teacher in your class.

Mitosis:

1) Take both your hands and hold them together. Your fingers represent the chromosomes in **prophase**.
2) Line your fingers up next to each other, interlocking with the fingers of your left-hand touch the corresponding fingers with your right hand. Now you are modeling **metaphase** with the chromosomes lined up in a row.
3) Now separate your hands; as you do this, you are modeling **anaphase**.
4) Now make two fists. Each hand becomes a nucleus in two cells during **telophase**.

Cytokinesis:

1) Form a circle with both hands by putting your fingertips and thumbs together; this represents one cell.
2) Slowly bring your fingertips down to your thumbs, eventually bringing them together, pinching the circle into two circles; this shows how the cytoplasm of one cell separates into two cells during cell division.

Questions:

1) How do these models show the two processes in cell division?

2) How are these models not accurate?

Cell Cycle Timing (With Slides)

Directions:

You will need **prepared slides of Onion Root Tips** and **Whitefish Blastula**, a **compound light microscope**, and **lens wipes** to study mitosis stages. **Looking at the materials and lab we will be using, what are the safety precautions we should take to protect ourselves and materials during the investigation?**

1) On each slide, compare how many cells you can see in each cell division stage by counting them and writing the data in Data Table 1.

Data Table 1

Phase	# of Onion Root Tip cells in each phase	# of Whitefish Blastula cells in each phase
Interphase		
Prophase		
Metaphase		
Anaphase		
Telophase		

Questions:

1) Which phase did you see has the most cells?

2) Was the data similar for both the Onion and Whitefish?

3) Why do you think some phases showed more cells and others showed fewer cells?

4) Of the phases, which do you think takes the longest time to complete?

5) Which phase(s) do you think takes the least amount of time to complete?

6) What could be some sources of error in this investigation?

Cell Cycle Timing (With Pictures)

Directions:

1) From each picture on page 190, count how many cells you can see in each cell division stage. Put this data in Data Table 1.

2) Then determine the percentage of time each cell spends at each stage of cell division. Divide each cell's number by the total number of cells and multiply by 100 to determine the percentage. Write this data in Data Table 1. Graph the percentages in Graph 1.

3) If these cells take 40 hours to go through their whole cycle, use the percentages to determine how much time is spent in each phase.

Data Table 1

	Plant 1	Plant 2	Plant 3	Total Cells	% of Total Cells Counted	Time in Each Phase
Interphase						
Prophase						
Metaphase						
Anaphase						
Telophase						
Total Cells	20	35	40	95	100%	40 hrs

Graph 1

Interphase	Prophase	Metaphase	Anaphase	Telophase

Pictures of Plant Cells

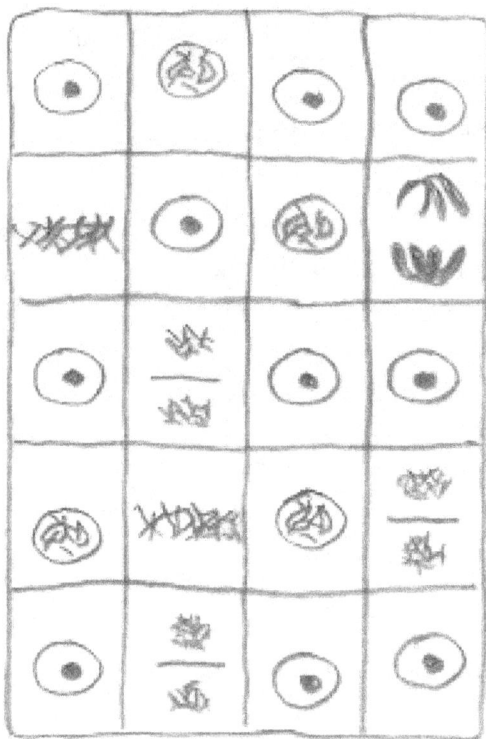

Plant 1

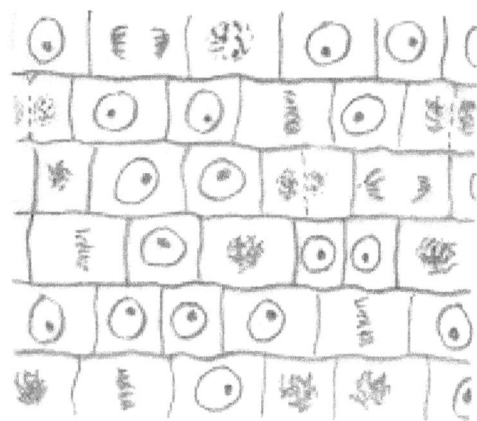

Plant 2

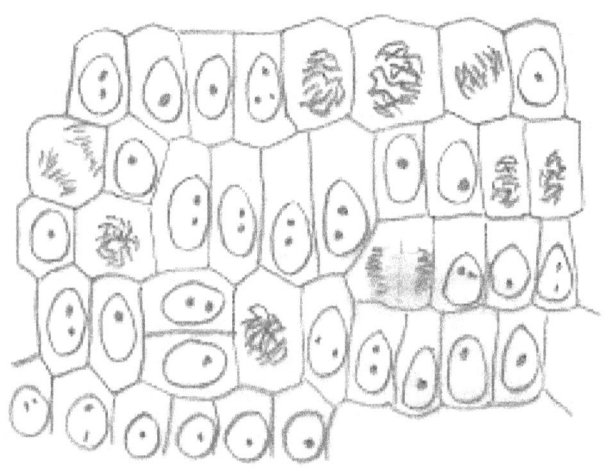

Plant 3

Questions:

1) Which phase of cell division takes the longest time? Explain how you know.

2) Which phase of mitosis takes the longest time? Explain how you know.

3) Which of the stages takes the shortest time? Why do you think that is?

Virtual Investigations that go with Unicellular Eukaryotes

ExploreLearning.com

Cell Structure Gizmo

Cell Energy Cycle Gizmo

Paramecium Homeostasis Gizmo

Cell Division Gizmo

Unit 6 Genetics

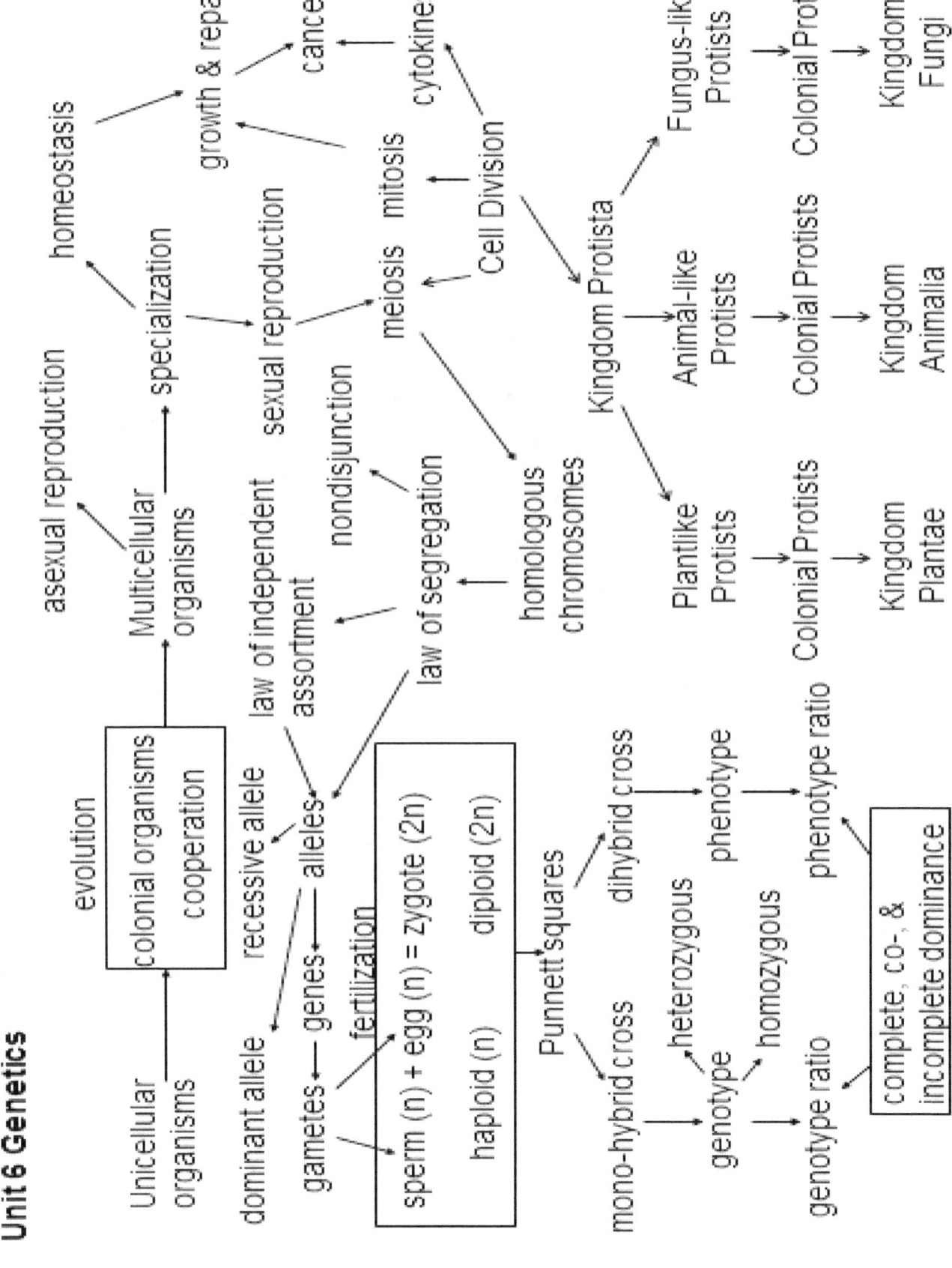

Unit 6 Compare and Contrast

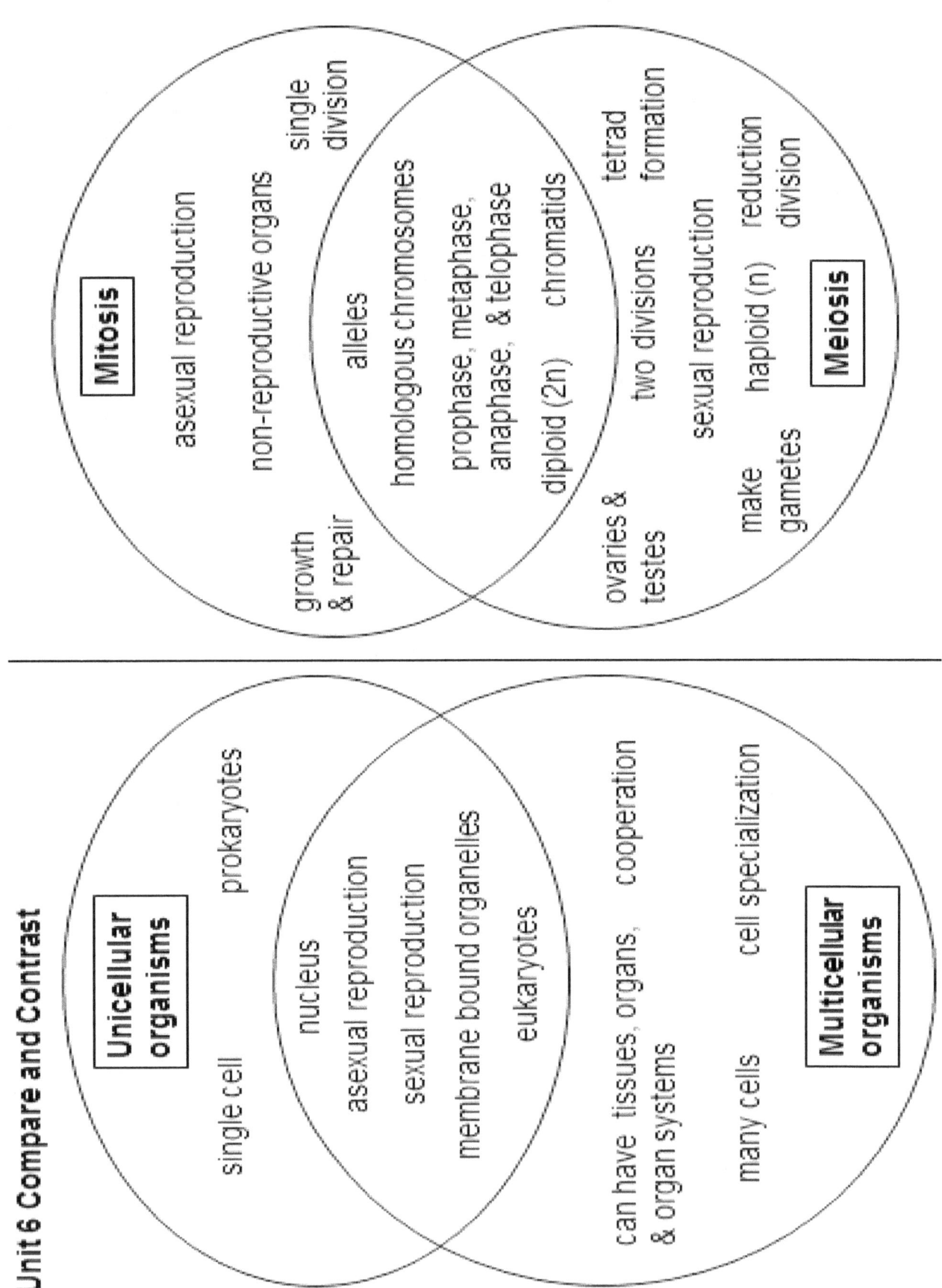

Top Venn diagram — **Mitosis** / **Meiosis**:

Mitosis only:
- asexual reproduction
- non-reproductive organs
- single division
- growth & repair

Overlap:
- alleles
- homologous chromosomes
- prophase, metaphase, anaphase, & telophase
- diploid (2n)
- chromatids

Meiosis only:
- tetrad formation
- two divisions
- reduction division
- sexual reproduction
- haploid (n)
- make gametes
- ovaries & testes

Bottom Venn diagram — **Unicellular organisms** / **Multicellular organisms**:

Unicellular organisms only:
- prokaryotes
- single cell

Overlap:
- nucleus
- asexual reproduction
- sexual reproduction
- membrane bound organelles
- eukaryotes

Multicellular organisms only:
- cooperation
- cell specialization
- can have tissues, organs, & organ systems
- many cells

Modeling Meiosis

Directions and Questions:

You will need two different color sets of **pipe cleaners**. Four pieces need to be cut to lengths of 1 inch, 2 inches, 3 inches, and 4 inches, two for each color. **Looking at the materials and lab we will be using, what are the safety precautions we can take to protect ourselves and materials during the investigation?**

1) Each two pipe cleaners that are cut to the same length and are the same color can be twisted together to make one chromosome with two chromatids.

2) Each pair of chromosomes (one of each color) represents a homologous pair of chromosomes.

3) **Prophase I**: Make a pile of all your chromosomes. There should be 8 of them. DNA has wound up into chromosomes. Four of one color and four of another. Each color represents the chromosomes given to you by each of your parents. This cell is **diploid (di-** meaning two**)**; it has double the chromosomes. How many chromosomes do you have in your pile?

      a. The first cell in males is called a **spermatocyte.**
      b. The first cell in a female is called an **oocyte.**

4) **Metaphase I**: Take your homologous pairs and line them up with each other, randomly placing different colors on opposite sides; this is how **independent assortment** occurs, lining homologous chromosomes up randomly.

5) **Anaphase I**: Separate your homologous pairs to opposite sides, so you have two lines of chromosomes.

6) **Telophase I**: Now, make piles out of your separated chromosomes. You have made **haploid** cells because they have half the chromosomes than they had before. At the end of this phase, the chromosomes unwind, and two nuclei form. Now we are halfway done. How many chromosomes do you have in each of your piles?

7) **Prophase II**: You already have Prophase II, two piles of chromosomes that wound back up.

8) **Metaphase II**: Take each of your piles and line up the chromosomes in them.

9) **Anaphase II**: Untwist each of the chromosomes, separating their chromatids. You should now have four lines of chromosomes.

10) **Telophase II**: Now, take your four lines of chromosomes and make them each into their own pile. These will end up being four different cells still being **haploid**. How many chromosomes are in each of your piles now?

   a. In males, all four will become **sperm** cells.
   b. In females, three of the cells will become small useless **polar bodies,** and one cell will become a larger **egg**.

11) **Fertilization**: This can be done by taking one of your four piles and combining them with another person's pile. This new cell would be called a **zygote**. This new cell/organism is now **diploid** because it has double the chromosomes from what it had before. How many chromosomes do you have now?

Comparing Expected Ratios to Experimental Ratios for a Monohybrid Cross

Directions:

You will need a separate **sheet of paper** and two **pennies. Looking at the materials and lab we will be using, what are the safety precautions we can take to protect ourselves and materials during the investigation?**

1) We will say the penny's head's side is dominant for hair (H), and the tail's side is recessive for having no hair (h). On a separate sheet of paper, construct a Punnett square to find the expected genotype and phenotype ratios; write them in Data Table 1.

2) Create a tally sheet on that separate sheet of paper to mark how many HH, Hh, and hh happen when you toss/flip the coins. Put the totals in Data Table 1 for 10, 100, and 1000 tosses.

Data Table 1

# of Tosses	# of HH	# of Hh	# of hh hairless	Total hairy (HH+Hh)	Expected Genotype Ratio	Expected Phenotype Ratio	Experimental Genotypic Ratio	Experimental Phenotypic Ratio
10								
100								
1000								

Questions:

1) Which phonotype happened more, hairy or hairless? Explain why.

2) Which genotype appeared the most?

3) Why do you think this genotype appeared so often?

4) Which phenotype appeared the least?

5) Why do you think this phenotype appeared the least?

6) How did the expected ratios compare with the experimental ratios?

7) How many experimental tosses produced a ratio closest to the expected ratios?

8) Which type of ratio shows the probability that something will happen?

9) Which type of ratio shows what actually did happen?

10) Will the expected ratios always match the experimental? Explain why.

11) How do flipping coins model independent assortment?

12) Why is independent assortment a law?

Paper Mates

Directions:

You will need two **pennies**. Heads represent a dominant trait, and tails represent a recessive trait. The parents are heterozygous for all four traits shown below. Flip two coins to see what each trait will be in each of the three offspring you create. Write the results in # 2. **Looking at the materials and lab we will be using, what are the safety precautions we should take to protect ourselves and materials during the investigation?**

Trait	Genotypes and Phenotypes	
Eyes	EE or Ee	ee
Nose	NN or Nn	nn
Teeth	TT or Tt	tt
Hair	HH Hh	hh

1) **Parents (P Generation):** Draw the parents below. They are heterozygous for all traits.

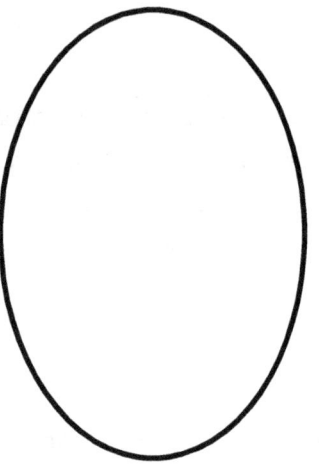 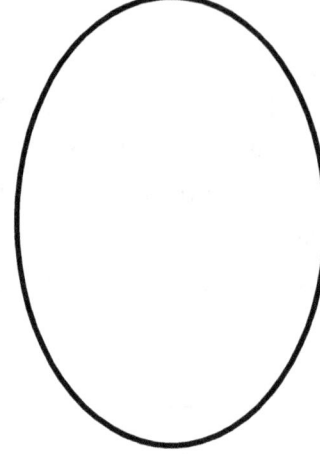

2) **Family (F1 Generation):** Write down the genotypes of each trait that happened from flipping the coins for each offspring. Then draw the three offspring that came from the flipping of the coins.

Genotype: Genotype: Genotype:

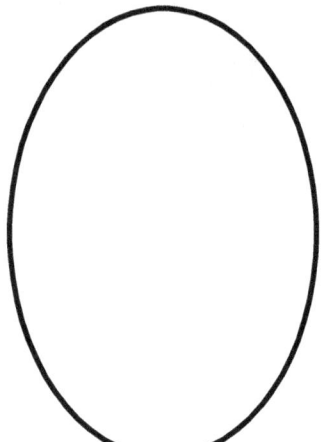 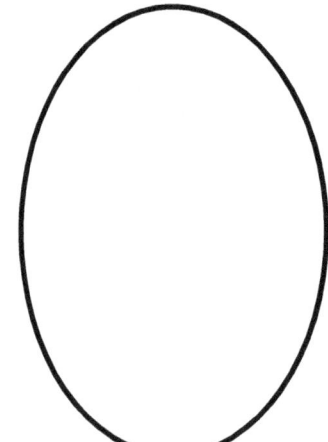 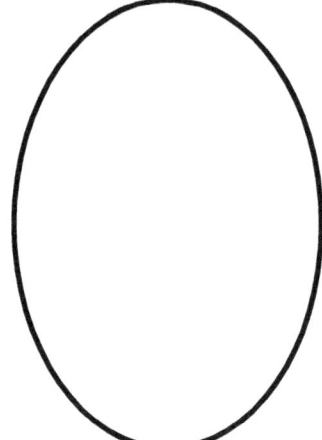

Create a Baby

Directions:

You will need one **penny** for each parent. One of you will be the mother, and one will be the father. **Looking at the materials and lab we will be using, what are the safety precautions we should take to protect ourselves and materials during the investigation?**

1) Begin with determining the sex of the baby. The mother can only pass on the female chromosome (X). The father can pass on a female (X) or a male chromosome (Y). Have the father flip the coin. Heads, the baby is male, tails, it is a female. Record this on Data Table 1.

2) Take turns flipping the coin to determine your baby's traits, starting with the head shape; heads will always be dominant, and tails will always be recessive. Fill in this information in Data Table 1. (**For example:** if the mother's coin lands on tails, she will pass on the recessive trait (r) for a round face. If the father's coin lands on heads, he will pass on the dominant trait (R) for an oval face. The baby's genotype will be (Rr), and its phenotype will be round-faced.) See page 203 for reference.

3) Repeat this information for each trait in Data Table 1. Use your own traits as the mother and father's genotype for eye color and hair color. (**For example:** if the mother has red hair, enter AAbb as the mother's genotype. If the father has dark brown hair, enter AaBB as the father's genotype. The mother would not have to flip since she can only give an Ab. The father would only flip for the "Aa"; the other gene will be B. So the baby will either be AABb or AaBb, depending on the flip.)

4) Draw your baby's face on page 204 based on the baby's phenotype on Data Table 1 and how it looks on the Genotype/Phenotype Reference Sheet on page 203.

Data Table 1

Trait	Mother's Genotype	Mother's Coin Flip	Father's Genotype	Father's Coin Flip	Baby's Genotype	Baby's Phenotype
Gender	XX	X	XY		X	
Face Shape	Rr		Rr			
Chin Shape	Nn		Nn			
Freckles	Ff		Ff			

Dimples	Dd		Dd			
Lip Thickness	Tt		Tt			
Eye Brows	Bb		Bb			
Eyelash	Ll		Ll			
Ear Lobes	Ee		Ee			
Widow's Peak	Ww		Ww			
Hair Curliness	Cc		Cc			
Nose Size	Ss		Ss			
Hair Color						
Eye Color						

Questions:

1) Are all traits only dominant and recessive?

2) Did you have a boy or a girl?

3) How do hair and eye color genetics differ from the other genes in this activity?

4) How are nose size and hair curliness different from the other traits?

5) How is homozygous freckled different from heterozygous freckled?

Trait	Homozygous Dominant	Heterozygous	Homozygous Recessive
Face Shape	RR Round	Rr Round	rr Oval
Chin Shape	NN Noticeable	Nn Noticeable	nn Less Noticeable
Freckles	FF Present	Ff Present	ff Absent
Dimples	DD Present	Dd Present	dd Absent
Lip Thickness	TT Thick	Tt Thick	tt Thin
Eye Brows	BB Bushy	Bb Bushy	bb Fine
Eye Lashes	LL Long	Ll Long	ll Short
Ear Lobes	EE Free	Ee Free	ee Free
Widow's Peak	WW Present	Ww Present	ww Absent
Hair Curliness	CC Curly	Cc Wavy	cc Strait
Nose Size	SS Small	Ss Medium	ss Large
Hair Color	AABB=Black AABb=Black AAbb=Red	AaBB=Brown AaBb=Brown Aabb=Blond	aaBB=Blond aaBb=Blond aabb=white (albino)
Eye Color	AABB=Deep Brown AABb=Deep Brown AAbb=Brown	AaBB=Green-Brown AaBb=Brown Aabb=Grey-Blue	aaBB=Green aaBb=Light Blue aabb=Pink (albino)

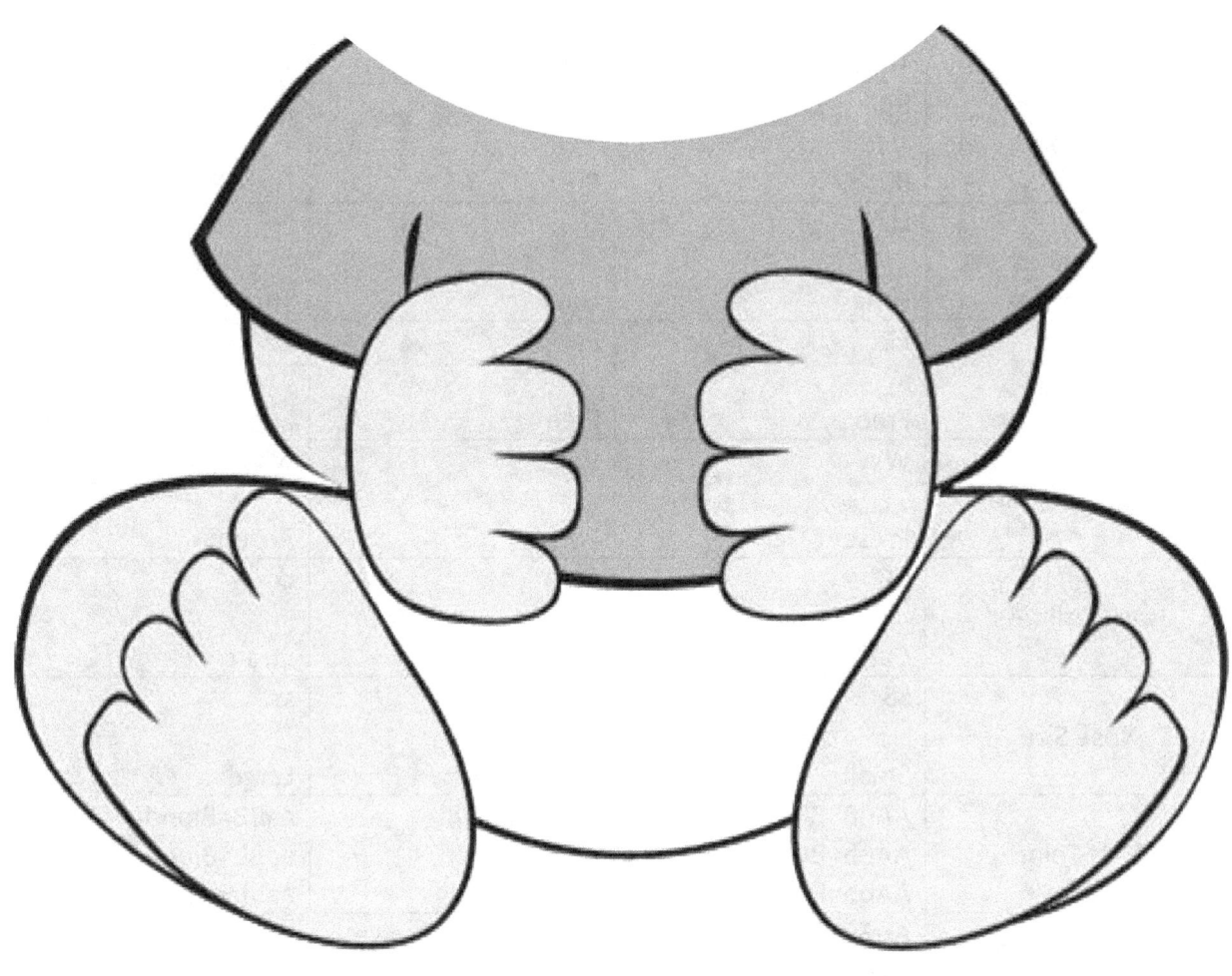

Construct Your Family Pedigree

Directions:

You will need to know as many relatives as possible to trace how you are related to them. Squares represent males, and circles represent females. Horizontal lines connecting to each other represent mating—vertical lines going down off the horizontal lines to represent births of offspring. Start at the bottom and work backward from yourself to see how far back you can trace your living and nonliving relatives.

Virtual Investigations that go with Genetics

ExploreLearning.com

Inheritance Gizmo

Chicken Genetics Gizmo

Mouse Genetics (One Trait) Gizmo

Mouse Genetics (Two Traits) Gizmo

Hardy-Weinberg Equilibrium Gizmo

Fast Pants 1 – Growth and Genetics Gizmo

Fast Plants 2 – Mystery Parent Gizmo

Meowsis STEM Case Gizmo

Meowsis Handbook Gizmo

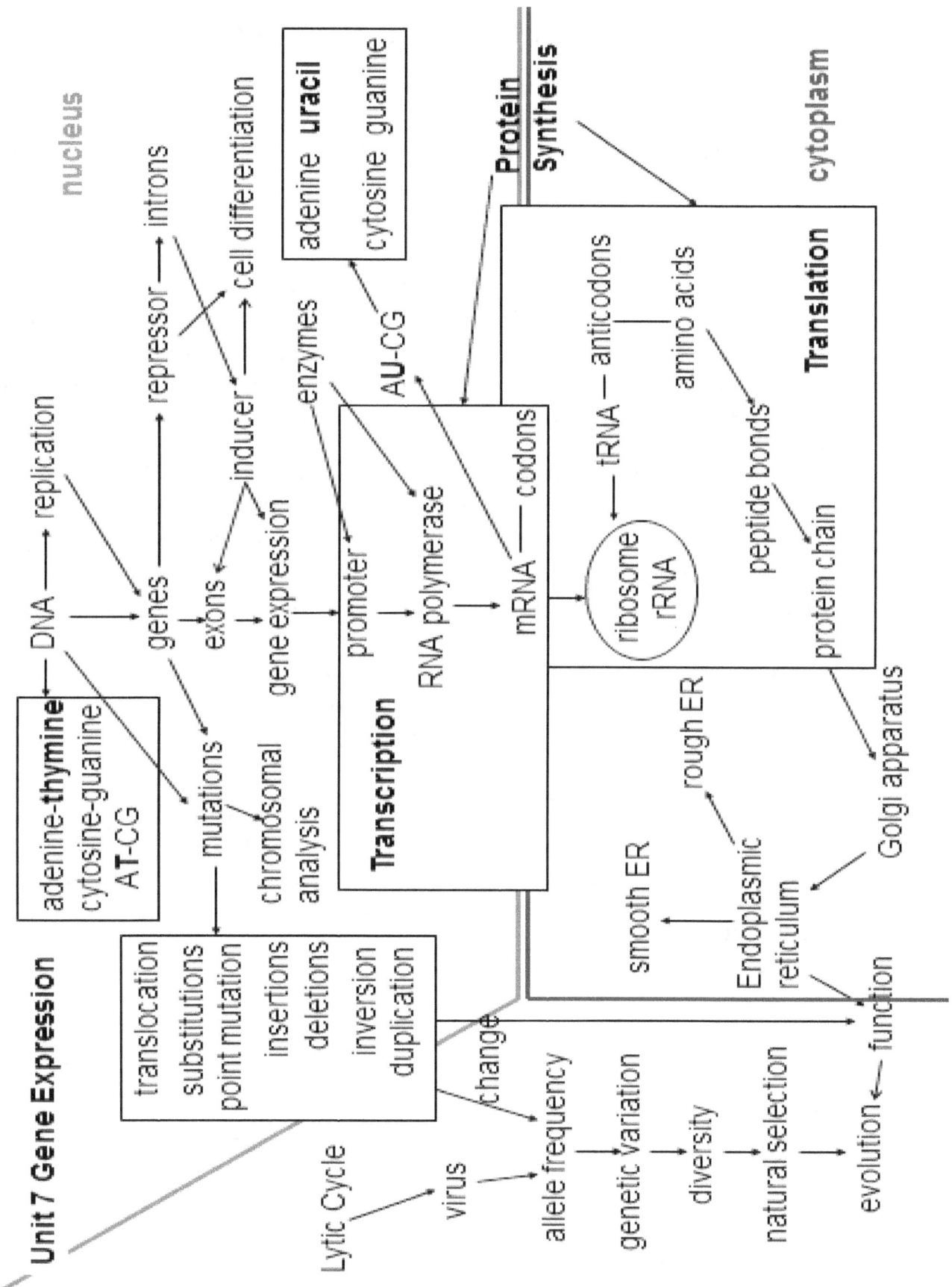

Protein Synthesis of the Quaddie

Directions:

You will need **colored pencils** to draw the Martians we did not recently find on Mars. Looking at their genetic code, we can see their weird protein structure that acts differently on Mars than on Earth. We suspect that some large collision or explosion shot microscopic organisms out of the Earth at some point in its past, sending them to Mars. Mutations that occurred along with the different forces of gravity and type of atmosphere caused this organism to develop differently and evolve to live in the conditions of Mars. Strangely mutations have taken place to cause the genes to shorten themselves to four codons long, which is extremely short compared to all organisms on Earth. Because these organisms only have eight genes, each with four codons, we call them Quaddies. (Remember code: **DNA: A-T**, **C-G**; **RNA: A-U, C-G**)

1) Use the DNA code in Data Table 1 for each gene to configure the mRNA codons. Then use those mRNA codons to see which tRNA would connect to the amino acids.

2) Then use the mRNA again with the Codon Table on the top of page 210 to find the amino acid sequence.

3) Once you have the amino acid sequences, you can use these to code for the organism's traits in Data Table 2 to fill in Data Table 1.

4) Use these traits to draw what you think this Quaddie looks like in Picture 1 on the bottom of page 210. Make sure you do not draw things that do not have a code.

Data Table 1

Gene 1	Gene 2
DNA: TAC AGA CTT CTG	DNA: GTA ATG TTT GGA
mRNA: _AUG_____	mRNA: _____
tRNA: _UAC_____	tRNA: _____
Amino Acids: _Methionine_____	Amino Acids: _____
_____	_____
Trait: _____	Trait: _____
Gene 3	**Gene 4**
DNA: CAT GCC TCA CCC	DNA: CGA GCT AAA TGA
mRNA: _____	mRNA: _____
tRNA: _____	tRNA: _____
Amino Acids: _____	Amino Acids: _____
_____	_____
Trait:_____	Trait: _____

Gene 5	**Gene 6**
DNA: TAA CGT ATA GCT	DNA: CAC TGC TTA AGG
mRNA: _____	mRNA: _____
tRNA: _____	tRNA: _____
Amino Acids: _____	Amino Acids: _____
_____	_____
Trait: _____	Trait: _____
Gene 7	**Gene 8**
DNA: ACC GTC ACA CCA	DNA: AAG TTG CTC ATC
mRNA: _____	mRNA: _____
tRNA: _____	tRNA: _____
Amino Acids: _____	Amino Acids: _____
_____	_____
Trait: _____	Trait: _____

Data Table 2

Amino Acid Sequence	Trait
Isoleucine-Alanine-Tyrosine-Arginine	Eyespot
Isoleucine-Alanine-Phenylalanine-Threonine	No eyespot
Aspartic Acid-Arginine-Serine-Cysteine	Hairy
Valine-Arginine-Serine-Glycine	No hair
Valine-Threonine-Asparagine-Serine	Fantail
Histidine-Threonine-Asparagine-Serine	Skinny tail
Methionine-Serine-Glutamine-Aspartic Acid	4 legs
Methionine-Serine-Glutamic Acid-Aspartic Acid	8 legs
Phenylalanine-Asparagine-Glutamic Acid-Stop	4 nostrils
Phenylalanine-Alanine-Glutamic Acid-Stop	2 nostrils
Alanine-Leucine-Phenylalanine-Tryptophan	No antennae
Alanine-Arginine-Phenylalanine-Threonine	4 antennae
Histidine-Tyrosine-Lysine-Proline	Green skin
Proline-Tyrosine-Lysine-Histidine	Red skin
Glycine-Glutamine-Cysteine-Tryptophan	Green spots
Tryptophan-Glutamine-Cysteine-Glycine	Red spots

Codon Table

		U	C	A	G	
U		Phenylalanine	Serine	Tyrosine	Cysteine	U
		Phenylalanine	Serine	Tyrosine	Cysteine	C
		Leucine	Serine	Stop	Stop	A
		Leucine	Serine	Stop	Tryptophan	G
C		Leucine	Proline	Histidine	Arginine	U
		Leucine	Proline	Histidine	Arginine	C
		Leucine	Proline	Glutamine	Arginine	A
		Leucine	Proline	Glutamine	Arginine	G
A		Isoleucine	Threonine	Asparagine	Serine	U
		Isoleucine	Threonine	Asparagine	Serine	C
		Isoleucine	Threonine	Lysine	Arginine	A
		Methionine	Threonine	Lysine	Arginine	G
G		Valine	Alanine	Aspartic acid	Glycine	U
		Valine	Alanine	Aspartic acid	Glycine	C
		Valine	Alanine	Glutamic acid	Glycine	A
		Valine	Alanine	Glutamic acid	Glycine	G
		U	C	A	G	

First Base (left axis) — Second Base (bottom axis) — Third Base (right axis)

Picture 1

Questions:

1) What is the difference between the DNA codons and mRNA codons?

2) What do you think is the amino acid Methionine's function if it is always used at the beginning of the gene code in real life?

3) What is the function of the Stop codon?

4) How is this organism genetically different from those found on Earth?

5) What part of this activity is a part of transcription?

6) What part of this activity is part of translation?

7) How are mRNA codons different from tRNA anticodons?

8) What if we know the amino acid sequence of a gene? How could we find the DNA code for that same gene?

9) Is there only one DNA code for that one gene? Or could multiple DNA codes code for the same amino acid sequence? Explain.

10) What has to change for a protein to change, the DNA or amino acid? Explain.

11) How is this model different from real life?

Protein Synthesis Role Play

Directions:

You will need a **box** (to be the nucleus), two **twisted phone cords** (to be the DNA molecule), **masking tape** (with a series of codons written on it to act as mRNA), **connecting alphabet baby letters** or **shapes** (to be the amino acids), **students** will be the tRNA (that bring the amino acids to the ribosome), a **plastic Walmart bag**, and your **teacher** and a **chair** (will act as the ribosome). **Looking at the materials and lab we will be using, what are the safety precautions we should take to protect ourselves and materials during the investigation?**

Construction Pre-Lab:

1) **Nucleus:** On the box, label the nucleus for everyone to see.
2) **DNA:** take two twisted phone cords and push them together to build the two strands of spiral DNA. Put the DNA in the nucleus box.
3) **mRNA:** On the masking tape, write a series of mRNA codons (starting with **AUG**) that do not repeat themselves but are in a random order for as many baby letters as you have to represent the 64 available codons. Carefully place another strand of tape on the back to cover the sticky back of the tape up. Place the mRNA inside the nucleus box.
4) **Amino Acids:** On the connecting baby toys that can build a long chain, write the complementary tRNA anticodons on each piece that go with all the mRNA codons.
5) **tRNA:** Randomly distribute all the baby toys with the tRNA anticodon on them to all the students in the class. Try to have everyone get the same amount.
6) **Ribosome:** place a chair for the teacher to sit in the middle of the room.

Modeling:

7) **Transcription:** The teacher will go to the nucleus box, unwind some of the phone cord DNA, and pull the masking tape mRNA from where the DNA splits to show transcription occurs there.
8) Pull all the mRNA tape out of the nucleus and put the DNA back together. Transcription is now over.
9) The teacher will now take the mRNA tape and sit in the chair to put the ribosome together. The teacher is one half of the ribosome, and the chair is the other.
10) **Translation:** The ribosome teacher sitting in the chair (the teacher cannot get up from the chair in the model) now will read the first codon to the class. The student (tRNA) with the complementary anticodon on their amino acid toy piece will get up and take it

to the teacher (ribosome). The teacher (ribosome) will inspect the amino acid anticodon to ensure it is correct. The teacher will link the amino acid to the chain if it is correct. If it is incorrect, it will reject it by returning it to the student transfer RNA.

11) The procedure in # 10 will be repeated until the entire mRNA chain is read and all the amino acids are put into the protein chain.

12) Once the chain is completed, the teacher folds up the chain to show how it makes a complicated shape. Put the protein chain in the plastic bag (packaging it) to send away to do its function. If the chain breaks on accident, we can say that the protein was denatured by something and will not work properly for its function; it might also do something else.

Questions:

1) How did transcription take place?

2) What represented the DNA?

3) What represented the nucleus?

4) What represented the mRNA?

5) How did translation take place?

6) What represented the amino acid?

7) What represented the tRNA?

8) What represented the ribosome?

9) What was the function of the tRNA?

10) What were the functions of the ribosome?

11) What happened to the protein chain at the end?

12) Describe protein synthesis.

Making a Karyotype

Directions:

1) Use the key below to build an individual's karyotype with a disorder. On page 217, you will need to label the remaining unlabeled chromosomes from 1-22, and X or Y. Number 23 will be different if it is male, the same bigger chromosome pair if female.

2) Next, use **scissors** to cut them out and either **tape** or **glue** them to the blank chart on page 219 at the appropriate spots. Make sure the centromeres are lined up on the horizontal line.

3) Identify where the abnormality is and research the **internet** or your **textbook** to identify the disorder.

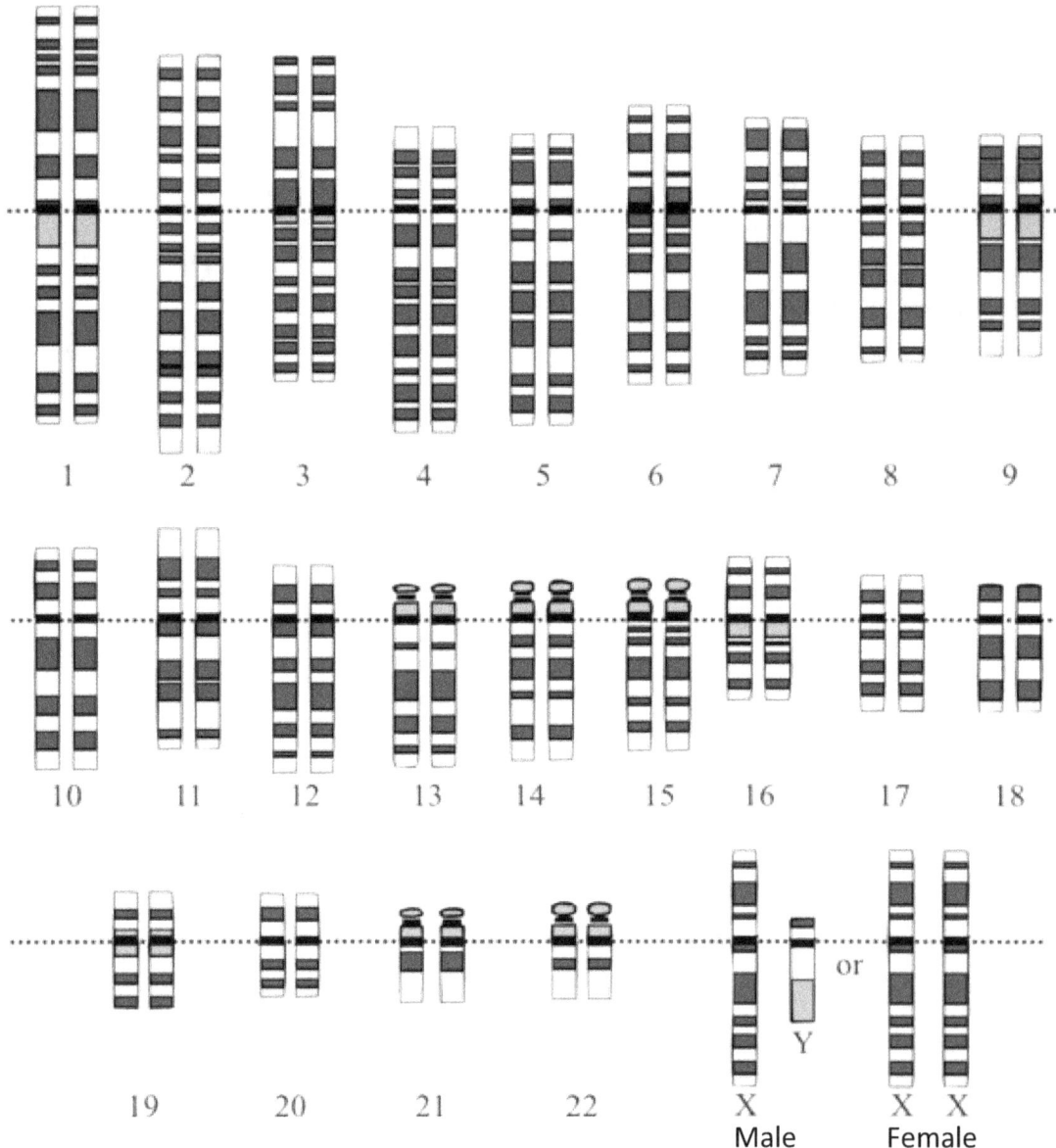

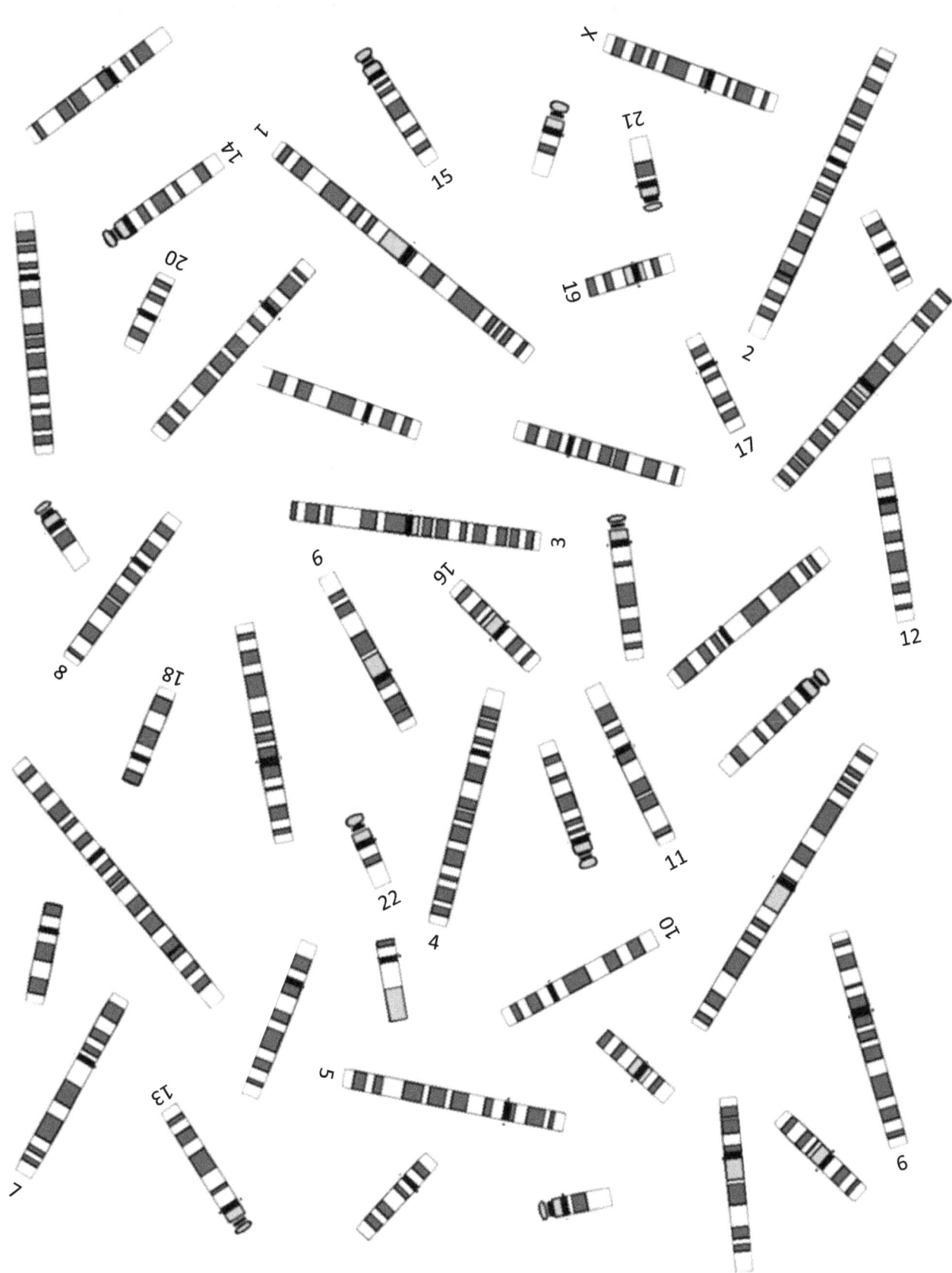

This page will be cut up from the page before!

Empty Chart

_____

1   2   3   4   5

_____

6   7   8   9   10   11   12

_____

13   14   15   16   17   18

_____

19   20   21   22   23

Questions:

1) How many sex chromosomes are in a normal human?

2) What is the difference between a normal male and a normal female on a karyotype?

3) Is your karyotype a male or female?

4) What did you observe was abnormal about your karyotype?

5) All the chromosomes that are not the sex chromosomes are the autosomes. How many autosomes were found in your karyotype?

6) How many autosomes are in a normal human karyotype?

7) What do you think would be worse, too many chromosomes or missing chromosomes? Explain.

8) How could you tell the different chromosomes apart?

9) How could you tell which chromosomes are homologous when building the karyotype?

10) Which phase of mitosis do you think was used to get the image of the chromosomes? Explain.

11) How do you think this investigation could be improved?

12) What kind of impact can this type of technology have on human lives and society?

13) What is the abnormality called that you found in your karyotype?

DNA Fingerprinting

Directions Part 1:

Look at the DNA Fingerprint below. A mother is accusing a man that he is the father of her child. Follow the directions below to help you analyze the autoradiographs of each of the three people involved to see if this is the birth father or not.

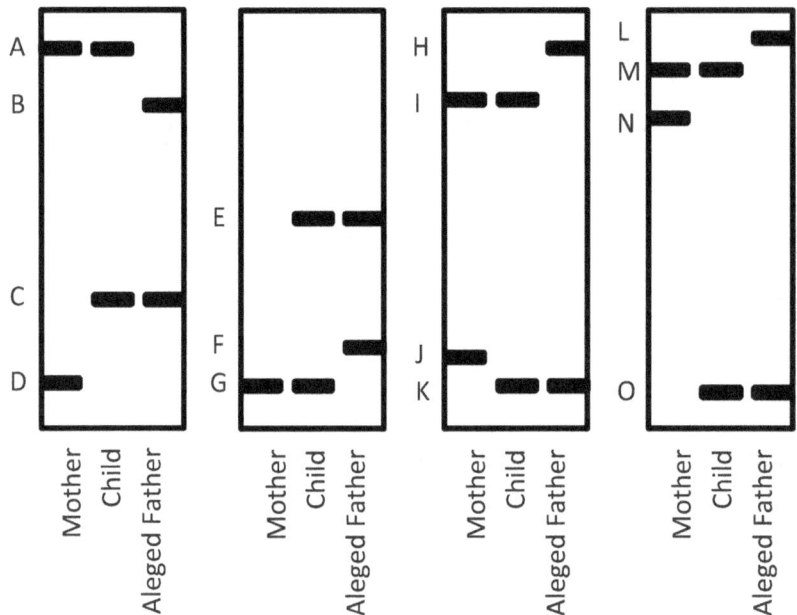

Questions Part 1:

1) Write the letters that are associated with each individual.
   a. Mother

   b. Child

   c. Alleged Father

2) Circle the genes of the child that came from the mother.

3) Circle the genes of the child that match up with the alleged father.

4) Are there any of the child's genes left over? What does this mean?

Directions Part 2:

Use the DNA fingerprints from the DNA taken from the crime scene and three suspects below to identify which suspect committed the crime.

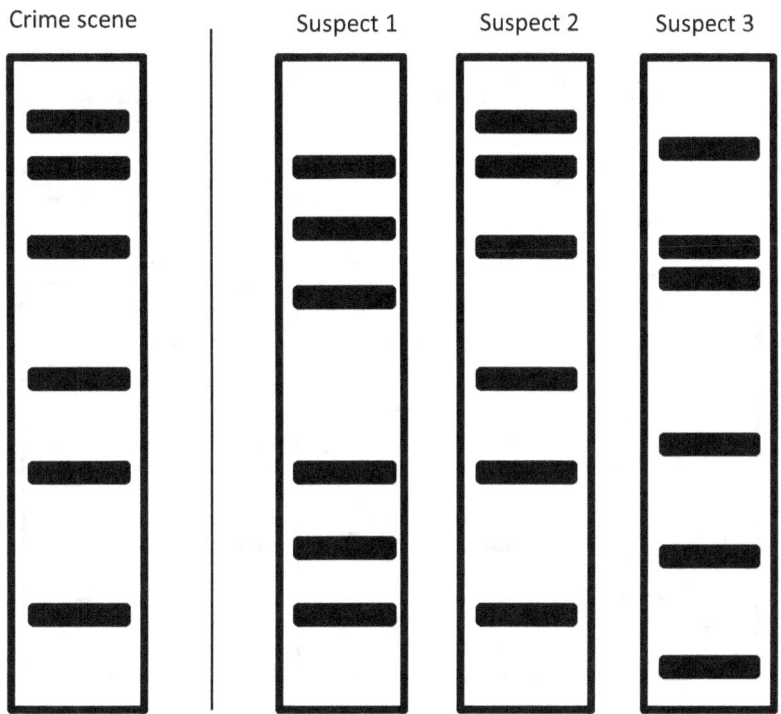

Questions Part 2:

1) How can you tell which suspect matches the DNA taken from the crime scene?

2) Which suspect is the criminal?

3) How do we know the other two suspects are not the criminals?

4) How else could DNA Fingerprints be used?

Directions Part 3:

Use a **ruler** to line up each of the babies' DNA marks with the three sets of parents. Each baby belongs to one set of parents that have been mixed up in the hospital. Match each baby with their parents, so each family goes home happy and whole.

Questions Part 3:

1) Which baby belongs to Couple A?

2) Which baby belongs to couple B?

3) Which baby belongs to couple C?

4) How did you know each baby belonged to the right couple?

5) How much information comes from each parent to make the baby?

Lego Mutation Models

Directions:

You will need a lot of **4 block Legos** of at least three different colors or a bag of **Mega Blocks** to build chromosome models simulating some of the different types of mutations. **Looking at the materials and lab we will be using, what are the safety precautions we can take to protect ourselves and materials during the investigation?**

1) Each chromosome is a pattern of information telling how to put together an organism. Use the images below and follow your teacher's instructions to stack your blocks with a pattern with two different colors to simulate a chromosome you will start with.
2) Produce a deletion mutation by taking a section out of your chromosome and connecting it back together.
3) Produce a duplication mutation by adding a pattern already in the chromosome.
4) Produce an inversion mutation by reversing the order of a section in your stack.
5) When simulating an insertion, use a third color to make another stack so you can see the insertion genes.
6) When simulating the translocation mutation, make two stacks where the pieces of each chromosome swap places

Types of Mutations

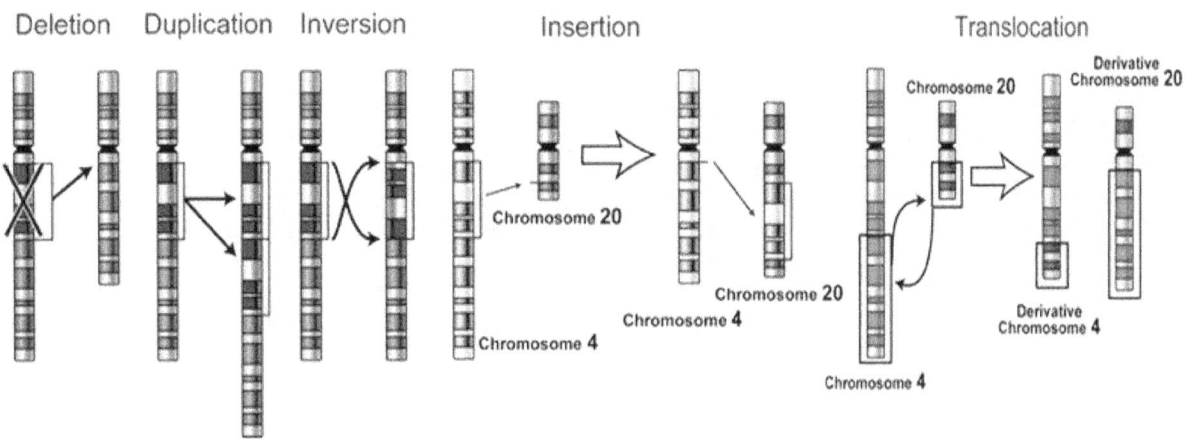

Questions:

1) What do you think could happen if you delete information from an organism?

2) What do you think could happen if you duplicate information in an organism?

3) What do you think could happen if you change the order of the information in an organism?

4) What do you think could happen if you add information to an organism?

5) How could each of these types of mutations benefit an organism?

   a. Deletion

   b. Duplication

   c. Inversion

   d. Insertion

   e. Translocation

6) How could each of these types of mutations harm an organism?

    a. Deletion

    b. Duplication

    c. Inversion

    d. Insertion

    e. Translocation

7) Which seems to be the most dangerous type of mutation you simulated today? Explain why.

8) How does mutation help the process of Natural Selection?

Lytic Virus Cycle

Directions:

Using the **internet** or your **textbook**, draw a virus lytic cycle showing its stages as it infects, reproduces, and spreads to other cells. Also, show how a lytic infection is different from a lysogenic infection.

Virtual Investigations that go with Gene Expression

ExploreLearning.com

RNA and Protein Synthesis Gizmo

Virus Lytic Cycle Gizmo

Human Karyotype Gizmo

DNA Analysis Gizmo

Protein Synthesis STEM Case Gizmo

Protein Synthesis Handbook Gizmo

PhET.colorado.edu

Gene Expression Essentials

Gene Machine: The Lac Operon

Unit 8 Plants

Photosynthesis

$$6H_2O + 6CO_2 \longrightarrow C_6H_{12}O_6 + 6O_2$$

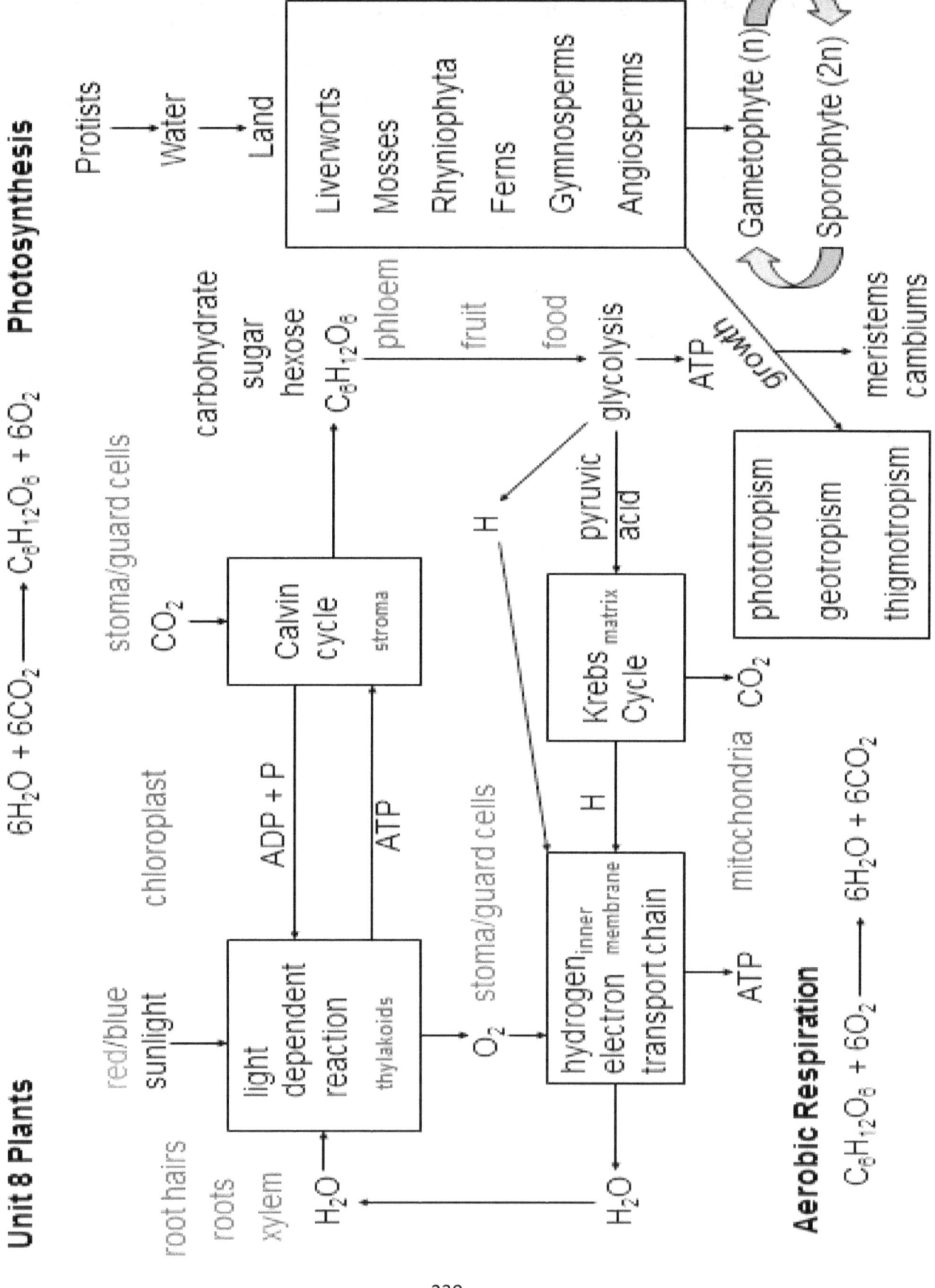

Aerobic Respiration

$$C_6H_{12}O_6 + 6O_2 \longrightarrow 6H_2O + 6CO_2$$

Evolution of Plants

Directions:

Use the **internet** and your **textbook** to research how plants evolved using the following guide. Use digital media of your teacher's choice to present this information.

1) What gave rise to the first multicellular plants?
2) When did **Liverworts** and **Mosses** evolve?
   a. Describe the structure of liverworts
   b. Provide pictures of some liverworts
   c. Describe or show how liverworts reproduce
   d. Describe the structure of mosses
   e. Provide pictures of mosses
   f. Describe or show how mosses reproduce
   g. Do they have vascular tissue?
3) When did **Rhynophyta** and **Ferns** evolve?
   a. Describe the structures of Rhynophyta
   b. Provide a picture of some Rhynophyta
   c. Describe how Rhynophyta reproduce
   d. Name some common members of Rhynophyta
   e. Describe the structure of ferns.
   f. Provide pictures of ferns.
   g. Describe how ferns reproduce?
   h. Name some common ferns
   i. Describe the vascular tissue of Rhynophyta and ferns.
4) When did **Gymnosperms** evolve?
   a. What does the word gymnosperm mean?
   b. Describe the structures of gymnosperms
   c. Provide pictures of some gymnosperms
   d. Describe how gymnosperm reproduce
   e. Describe the vascular tissue of gymnosperms
   f. Name some common gymnosperms
5) When did **Angiosperms** evolve?
   a. What does the word angiosperm mean?
   b. Describe the structures of angiosperms
   c. Provide pictures of some angiosperms
   d. Describe how angiosperms reproduce
   e. Describe the vascular tissue of angiosperms

       f.   Name some common Angiosperms

6) How do all these plants go through the **alternation of generations**?

7) What is the function of **meristems**?

       a.   Which plants have them?

8) What is the function of **cambiums**?

       a.   Which plants have them?

9) Provide a timeline of when the different types of plants evolved

Root Stem and Leaf Cross Sections

Directions and Questions:

You will need a **compound light microscope**, **lens paper**, and **prepared slides of cross-sections of monocot** and **dicot roots, stems, and leaves. Looking at the materials and lab we will be using, what are the safety precautions we should take to protect ourselves and materials during the investigation?**

1) **Monocot root:** Use your teacher's instructions, the skills you have practiced, and your experience to center, focus, and draw a monocot root under the compound light microscope. Make sure to include and label the vascular tissue (xylem and phloem) here:

2) **Dicot root:** Use your teacher's instructions, skills you have practiced, and your experience to center, focus, and draw a cross-section of a dicot root under the compound light microscope. Make sure to include and label the vascular tissue (xylem and phloem) here:

3) What visual differences did you see between the monocot and dicot root cross-sections?

4) **Monocot stem**: Use your teacher's instructions, the skills you have practiced, and your experience to center, focus, and draw a cross-section of a monocot stem under the compound light microscope. Make sure to include and label the vascular tissue (xylem and phloem) here:

5) **Dicot stem**: Use your teacher's instructions, the skills you have practiced, and your experience to center, focus, and draw a cross-section of a dicot stem under the compound light microscope. Make sure to include and label the vascular tissue (xylem and phloem) here:

6) What visual differences did you see between the monocot and dicot stem cross-sections?

7) **Monocot leaf**: Use your teacher's instructions, the skills you have practiced, and your experience to center, focus, and draw a cross-section of a monocot leaf under the compound light microscope. Make sure to include and label the vascular tissue (xylem and phloem) here:

8) **Dicot leaf**: Use your teacher's instructions, skills you have practiced, and your experience to center, focus, and draw a cross-section of a dicot leaf under the compound light microscope. Make sure to include and label the vascular tissue (xylem and phloem) here:

9) What visual differences did you see between the monocot and dicot leaf cross-sections?

10) What are some differences between monocots and dicots that you saw in the investigation today?

11) Use what you have just observed and learned to fill in Data Table 1 showing monocot and dicot characteristics.

Data Table 1

Plant Parts	Monocots	Dicots
Cross Section of Roots		
Cross Section of Stems		
Cross Section of Leaves		

Tap Roots vs. Fibrous Roots

Directions and Questions:

You will need **whole carrots** and **small onions** (used for chives). **Looking at the materials and lab we will be using, what are the safety precautions we should take to protect ourselves and materials during the investigation?**

1) Looking at the white tip of the onion, this is an example of a **fibrous root** system. Draw a picture and write some characteristics here that describe the onion roots.

2) What do you think are the reasons the fibrous root is structured like it is. Discuss with your teacher and classmates.

3) Looking at the carrot, this is an example of a **taproot**. Draw a picture and write some characteristics here that describe the carrot root.

4) What do you think are the reasons the taproot is structured like it is. Discuss with your teacher and classmates.

Monocot vs. Dicot

Directions and Questions:

You will need **leaf** and **flower samples** of monocots and dicots. **Looking at the materials and lab we will be using, what are the safety precautions we should take to protect ourselves and materials during the investigation?**

1) Looking at the monocot leaves, what patterns do you see in the leaves of the monocots?

2) Looking at the dicot leaves, what patterns do you see in the leaves of dicots?

3) How are the dicot leaves different from the monocot leaves?

4) Looking at the monocot flowers, what patterns do you see in the flowers of monocots?

5) Looking at the dicot flowers, what patterns do you see in the flowers of dicots?

6) How are the dicot flowers different than the monocot flowers?

7) Use what you have just observed and learned to fill in Data Table 1 showing monocot and dicot characteristics.

Data Table 1

Plant Parts	Monocots	Dicots
Characteristics of Leaves		
Characteristics of Flowers		

Germination and Tropisms Lab

Directions:

You will need a **Ziploc freezer bag**, **water**, a **pipette**, a **paper towel**, five dry **beans**, and a **stapler** or **tape**. **Looking at the materials and lab we will be using, what are the safety precautions we should take to protect ourselves and materials during this investigation?**

1) Make a fold in the paper towel to create a trough where the bean seeds can sit. Slide the paper towel into the freezer bag where the towel touches the bottom of the bag, and the trough sits at the top. Staple the bag and paper towel in a row just under the trough on the paper towel.

2) Staple the bag to the wall or if you have a window, tape it to the window so that the bottom of the bag sits on the window sill, supporting the bag so that it does not fall over.

3) Place five beans in the trough of the paper towel.

4) Use a pipet to quirt water into the bag until it lines the bottom of the bag. Make sure when the water soaks up the paper towel to the seeds, there is still water in the bottom of the bag. Check each day to ensure there is enough water in the set-up so that the paper towel near the seeds stays wet.

5) Each day, measure the length of the stems and roots in millimeters for two weeks. Fill in Data Table 1 on page 240. Cross out the days that are on the weekend.

Questions:

1) Which direction did all the roots grow?

2) Which direction did the stems grow?

3) What do you think "forced" this to happen? This process is called **gravitropism**.

4) Where does it look like the leaves are facing and growing? This process is called **phototropism**.

5) When you see a path where grass will not grow because it is being walked on each day, or the grass under your house's eve after a rainstorm where the grass does not grow. These are examples of **thigmotropism:** a reaction of plants to touch. Why do you think we prune dead branches off of bushes and trees?

Data Table 1

Day	Length of Stem 1 (mm)	Length of Root 1 (mm)	Length of Stem 2 (mm)	Length of Root 2 (mm)	Length of Stem 3 (mm)	Length of Root 3 (mm)	Length of Stem 4 (mm)	Length of Root 4 (mm)	Length of Stem 5 (mm)	Length of Root 5 (mm)
1										
2										
3										
4										
5										
6										
7										
8										
9										
10										
11										
12										
13										
14										

Conservation of Life: Photosynthesis and Respiration

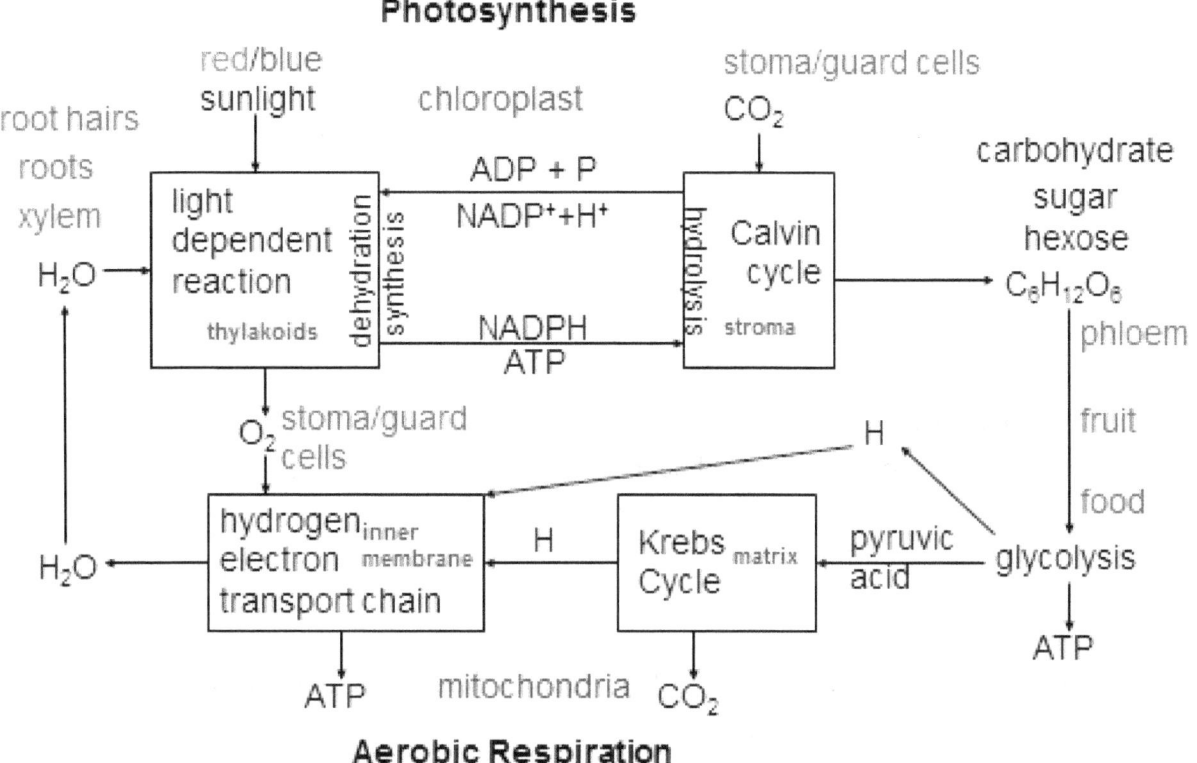

Directions:

Use the diagram above to answer the following questions.

1) Use the diagram above to write the equation for photosynthesis and balance the reaction.

2) Use the diagram above to write the equation for aerobic respiration and balance the reaction.

3) In both equations, trace where each element of the reactants go to make the products.

4) How are the two reactions similar to each other?

5) How does the conservation of mass, in this case, show that life has balance?

6) Plants and algae go through both photosynthesis and respiration. Animals only go through respiration. What would happen to life on Earth if we lost the plants and algae?

7) Keeping this in mind, which do you think formed first: the process of aerobic respiration or photosynthesis? Explain why.

8) How does the water needed for photosynthesis get into plants?

9) How does the carbon dioxide needed for photosynthesis get into plants?

10) Where does the glucose go in the plant?

     a. What will it be used for?

11) Where does the oxygen go?

      a.   What will it be used for?

12) Where in the cell does photosynthesis happen?

13) Where in the cell does aerobic respiration happen?

14) Both organelles are a form of bacteria that live symbiotically in plants; they even have their own circular DNA. Do plants go through respiration?

Building a Model of a Water Molecule

Directions:

You will need a **balloon**, a **molecular model kit,** and a **Periodic Table. Looking at the materials and lab we will be using, what are the safety precautions we should take to protect ourselves and materials during the investigation?**

1) At the top of your periodic table, label it like this just below:

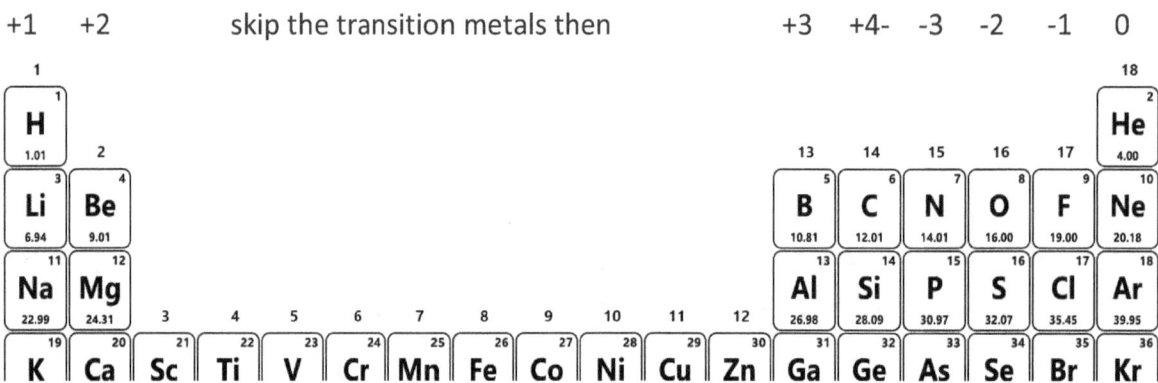

2) Different kits have different colors. In my kit, the:
   a. +1 (one-prong white) represents the Alkali Metals
   b. +2 (two-prong yellow) represents the Alkaline Earth Metals
   c. +3 (three-prong blue) represents the Boron Group
   d. +/- 4 (four-prong black) represents the Carbon Group
   e. -3 (three-prong red) represents the Nitrogen Group
   f. -2 (two-prong blue) represents the Oxygen Group
   g. -1 (one-prong green) represents the Halogens

3) Use the pieces to make two H_2O molecules. The hydrogen side of the molecule is slightly positive, and the oxygen side of the molecule is slightly negative making it polar like a magnet.

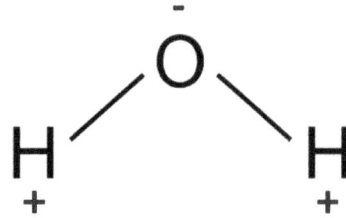

4) Because the water molecule has positive and negative ends, **ions** are attracted to the opposite charges on the water molecule. The positive ions are attracted to the oxygen side, and the negative ions are attracted to the hydrogen side. The same is true for other **polar molecules**; this is why ionic compounds and polar molecules like to dissolve in water. We call water the **universal solvent**.

5) Make a model of liquid water by taking your two water molecules and placing them next to each other where the oxygen of one is sitting between the two hydrogens of the other; this is how water molecules like to stick to each other. The positive ends are attracted to the negative ends; this is why water is **cohesive** (it sticks together).

6) When there are a bunch of them together, they have an equal pull on each other except for the ones on the surface; they are pulled slightly down because they have a slight charge above them. After all, there are no other molecules above them; this is why water has **surface tension**.

7) Since it is charged on both ends, it is also attracted to surfaces like a balloon with a static charge is attracted to a sweater or a wall. This attraction is why see water clinging to the sides of cold cans or glasses of ice tea. This phenomenon is called **adhesion**.
   a. You can model this by taking an inflated balloon, rubbing it on your hair to steal some electrons, and then sticking it to a shirt or wall.

8) You can make a model of ice (solid water) by flipping one of your water molecules and facing the oxygen ends toward each other. When water gets cold, the molecule's charge is not as strong, and the opposite ends are not attracted to each other anymore, so the oxygen atoms come together and share electrons with each other bonding them together. This orientation gives the molecule more space inside it and is why ice floats in liquid water.

Question:

1) Why do you think life depends so much on water?

Celery Transport

Directions and Questions:

You will need fresh **celery** with leaves, a **beaker**, and **red** and **blue food coloring**. This investigation will also work with a **white carnation** and **all colors of food coloring. Looking at the materials and lab we will be using, what are the safety precautions we should take to protect ourselves and materials during the investigation?**

1) In a beaker of water, add a couple of drops of food coloring. Put the celery into the water with the leaves up out of the water. What do you think will happen with the water in the beaker and the celery? **Hypothesis:**

2) Let the water and celery sit overnight. Come back the next couple of days and observe what you see. What did you see in the celery's leaves?

3) What do you think caused the celery to look like this?

4) How is the xylem in the celery stem like a straw in a drink?

5) Discuss with your teacher and class how pressure and magnetism were involved with this process you observed in this investigation. Explain what you discussed.

6) What do you think would happen if we split a white carnation stem into four sections and put each section into different colors of water? Test it and see.

Transpiration Pull

Directions and Questions:

You will need a **pressure sensor** and the **tube setup** attached to an **interface** connected to a **computer** with **Logger Pro**. You will then need to find a **plant branch** that will snuggly fit inside the tube, sealing the tube. **Looking at the material and lab we will be using, what are the safety precautions we should take to protect ourselves and materials during the investigation?**

1) Once your pressure sensor is connected to the interface and the computer with the Logger Pro, set the data collection to collect data for 5 minutes. Put the plant branch inside the tube connected to the pressure sensor, so it is sealed.

2) Press "Collect" on the Logger Pro. Watch the data for a few minutes. What do you see happening to the measurement of the pressure sensor?

3) Why do you think this is happening?

4) When the data seems to level out, break the seal, let the pressure equalize with the atmosphere, and stop the data collection if it has not already stopped. Set up the experiment to run again. Did you see the same trend in the results?

5) How does this show evidence of transpiration pull?

Seeing a Stoma

Directions and Questions:

You will need a **textbook, clear scotch tape, lettuce**, a **slide**, and a **compound light microscope**. **Looking at the materials and lab we will be using, what are the safety precautions we should take to protect ourselves and materials during the investigation?**

1) Take a small piece of scotch tape, put the sticky side on the lettuce, and then peel it off. You should have just removed one layer of cells from the outside of the lettuce. There you should see the epidermis, which contains the guard cells and stomata.

2) Place the sticky side of the tape down on the slide. Place the slide on the microscope stage and follow your teacher's instructions on centering and focusing it. Draw a picture of the lettuce epidermis labeling the guard cells and stoma.

3) Research how the guard cells open and close the stoma and describe it here:

4) Why does the rest of the epidermis look like puzzle pieces?

5) What is the function of the epidermis?

6) What is the function of a stoma?

Flower Dissection

Directions and Questions:

You will need a typical **flower** and a **scalpel. Looking at the materials and lab we will be using, what are the safety precautions we should take to protect ourselves and materials during the investigation?**

1) Obtain a flower from your teacher. Draw your flower in the space provided here. Note the color and label the sepals and petals.

2) Using the scalpel, very carefully make a vertical incision to open your flower if it is not already open. In the space provided here, draw your flower open. Be sure to label the **pistil** (female part) that has the sticky **stigma** on top; the shaft is called a **style,** the base contains the **ovary**, and the **stamen** (male part) that has an **anther** on the top of a **filament.**

3) Which is longer, the pistil or the stamen?

4) Is your flower set up to self-pollenate or be pollinated by a vector, like an insect, bat, or bird? Explain.

5) Do you see any powdery residue? This powder would be pollen (plant sperm) used to fertilize the eggs inside the ovary (and give us allergies).

6) What are the female parts of the flower?

7) What are the male parts of the flower?

8) Why would having a longer pistil or stamen be an advantage in pollination?

9) How many petals does your flower have?

10) Look at the veins on the leaves of your flower, are they branched or parallel?

11) Is your flower a monocot or a dicot?

Seed Dissection

Directions and Questions:

You will need **peanuts** with their shell on, **sunflower seeds** with their shell on, and a **paper towel. Looking at the materials and lab we will be using, what are the safety precautions we should take to protect ourselves and materials during the investigation?** (Make sure no one has an allergy to peanuts.)

1) On the paper towel, carefully open the peanut shell. Notice the brown to reddish-brown paper-like covering around the seed; this is the **seed coat** that protects the seed if it is swallowed whole.

2) How many sections do you see in the single seed?

3) These are the seed leaves and the reason why we call them a dicot. Split it in half carefully and notice the baby plant embryo sticks to one of the two cotyledons. Draw a picture of the embryo on the cotyledon and label the **cotyledon**, **hypocotyl** (embryo's leaves), **Epicotyl** (embryo's stem), and the **radicle** (the embryo's root) in the seed.

4) Now open the shell of the sunflower seed. Does it have two sections like the peanut?

5) Since there is only one seed leaf here, it is called a monocot. What do you think is in the cotyledons of the seeds that help the embryo grow before the sun hits it to allow photosynthesis?

6) If it is safe, and your teacher ok with it, you may now eat the seeds and clean up your area. The seeds taste good because energy tastes good, so we want to eat them.

Virtual Investigations that go with Plants

ExploreLearning.com

Photosynthesis Lab Gizmo

Plants and Snails Gizmo

Dehydration Synthesis Gizmo

Flower Pollination Gizmo

Pollination: Flower to Fruit Gizmo

Germination Gizmo

Seed Germination Gizmo

Growing Plants Gizmo

Measuring Trees Gizmo

Photosynthesis STEM Case Gizmo

Photosynthesis Handbook Gizmo

Cell Respiration STEM Case Gizmo

Cell Respiration Handbook Gizmo

Unit 9 Evolution of Animals

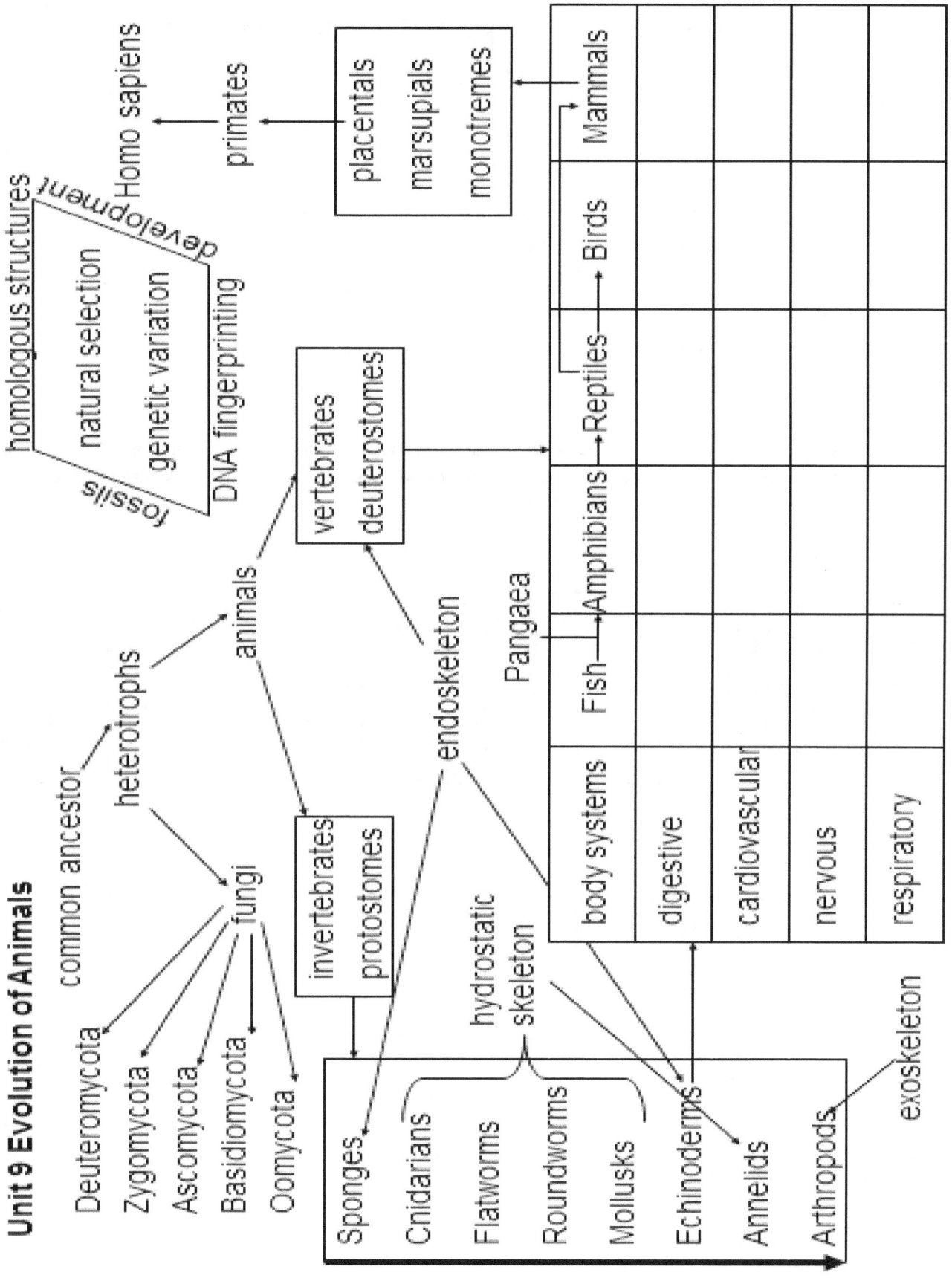

Evolutionary Relationships Seen Through Amino Acids

Directions:

1) Use the key below to compare amino acid sequences of different mammals to check for relatedness. The more similarities, the closer the relationship; the more differences, the less relationship. We will look at part of a protein that is common to all mammals, hemoglobin. Circle where the hemoglobin chain is different from the human in each of the vertebrates seen in Data Table 1. Then make a bar graph of the number of differences you found in Graph 1.

2) Then use the data in Data Table 2 to make a bar graph for Graph 2.

Key:

G-Glycine   A-Aspartic Acid   V-Valine   U-Glutamic Acid   L-Lysine   I-Isoleucine   P-Phenylalanine   M-Methionine   C-Cysteine   S-Serine   B-Glutamine   H-Histidine   T-Threonine   V-Valine   D-Proline   N-Asparagine   E-Leucine   F-Arginine   K-Tyrosine   O-Alanine   Q-Tryptophan

Data Table 1

Human	T	E	S	B	E	H	C	A	L	E	H	V	A	D	B	N	P	F	E	E	G	N	V	E	V	C	V	E	O	H
Chimp	T	E	S	B	E	H	C	A	L	E	H	V	A	D	B	N	P	F	E	E	G	N	V	E	V	C	V	E	O	H
Gorilla	T	E	S	B	E	H	C	A	L	E	H	V	A	D	B	N	P	L	E	E	G	N	V	E	V	C	V	E	O	H
Rhesus monkey	B	E	S	B	E	H	C	A	L	E	H	V	A	D	B	N	P	L	E	E	G	N	V	E	V	C	V	E	O	H
Horse	O	E	S	B	E	H	C	A	L	E	H	V	A	D	B	N	P	F	E	E	G	N	V	E	O	E	V	V	O	F
Kangaroo	L	E	S	B	E	H	C	A	L	E	H	V	A	D	B	N	P	L	E	E	G	N	I	I	V	I	C	E	O	B

Data Table 2

Human cytochrome c vs.	Number of Differences
Chimpanzee cytochrome c	0
Fruit fly cytochrome c	29
Horse cytochrome c	12
Pigeon cytochrome c	12
Rattlesnake cytochrome c	14
Red bread mold cytochrome c	48
Rhesus monkey cytochrome c	1
Screwworm fly cytochrome c	27
Snapping turtle cytochrome c	15
Tuna cytochrome c	21
Wheat cytochrome c	43

Graph 1 Number of Differences in Hemoglobin with Human

10									
9									
8									
7									
6									
5									
4									
3									
2									
1									

Chimp	Gorilla	Rhesus Monkey	Horse	Kangaroo

Graph 2 Number of Differences in Cytochrome c with Human

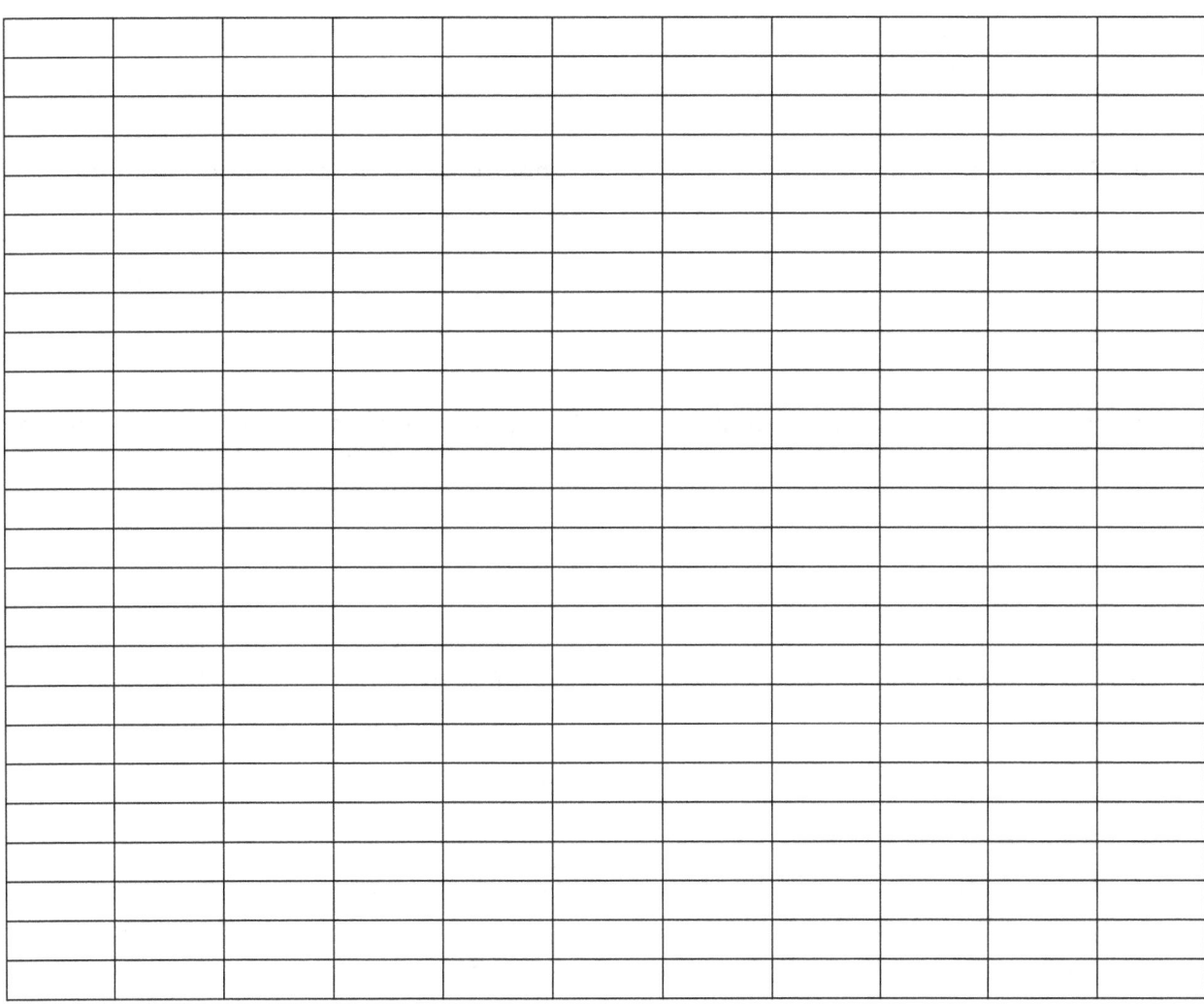

Chimp   Fr. Fly   Horse   Pigeon   R. Snake   RB Mold   Monkey   S. Fly   Turtle   Tuna   Wheat

Questions:

1) According to this data, which organisms are the most closely related to humans?

2) Which of these organisms are the least related to humans?

3) In Graph 1, explain why you think Kangaroos have more differences than the other mammals.

4) Primates are all in the same order as each other. How does this data show primates are more closely related to each other than other mammals?

5) What does this data show us about the relatedness of organisms with humans that are not animals?

6) Which organism probably has the most recent common ancestor with a human?

7) The fossil record says fish gave rise to amphibians, giving rise to reptiles, giving rise to birds and mammals. Mammals gave rise to primates, giving rise to apes (no tail), giving rise to hominids. How does the data in this investigation support the fossil record?

8) What do you think caused the changes in amino acids between these organisms?

9) If the amino acid sequence is similar between organisms, what does that say about their DNA? Explain.

10) Even though Red Bread Mold has 48 differences of 104 amino acids, they have 56 similarities. Explain why this is evidence that all life has a common ancestor with all other life?

11) How does this activity give evidence to the Theory of Evolution?

12) Why is evolution a theory?

Comparing Vertebrates

Directions:

Use the key below to compare amino acid sequences of different primates and non-primates to check for relatedness. The more similarities, the closer the relationship; the more differences, the less relationship. We will look at part of a protein that is common to all mammals, hemoglobin. Circle where the hemoglobin chain is different from the human in each of the vertebrates seen in Data Table 1. Then graph the number of similarities you found in Graph 1.

Key:

G-Glycine   A-Aspartic Acid   V-Valine   U-Glutamic Acid   L-Lysine   I-Isoleucine   P-Phenylalanine   M-Methionine   C-Cysteine   S-Serine   B-Glutamine   H-Histidine   T-Threonine   V-Valine   D-Proline   N-Asparagine   E-Leucine   F-Arginine   K-Tyrosine   O-Alanine   Q-Tryptophan

Data Table 1

Human	S	T	O	G	A	U	V	U	A	T	D	G	G	O	N	O	T	F	H
Chimp	S	T	O	G	A	U	V	U	A	T	D	G	G	O	N	O	T	F	H
Gorilla	S	T	O	G	A	U	V	U	A	T	D	G	G	O	N	O	T	L	H
Baboon	N	T	T	G	A	U	V	N	A	S	D	G	G	N	N	O	G	L	H
Lemur	O	T	O	G	U	L	V	U	A	S	D	G	S	H	N	O	G	L	H
Dog	S	S	G	G	A	U	I	N	A	T	D	S	N	L	N	O	O	L	L
Chicken	B	T	G	G	O	U	I	A	N	S	D	B	T	L	N	S	B	F	O
Frog	A	S	G	G	L	H	V	V	N	S	O	H	O	L	N	O	L	F	F

Figure 1

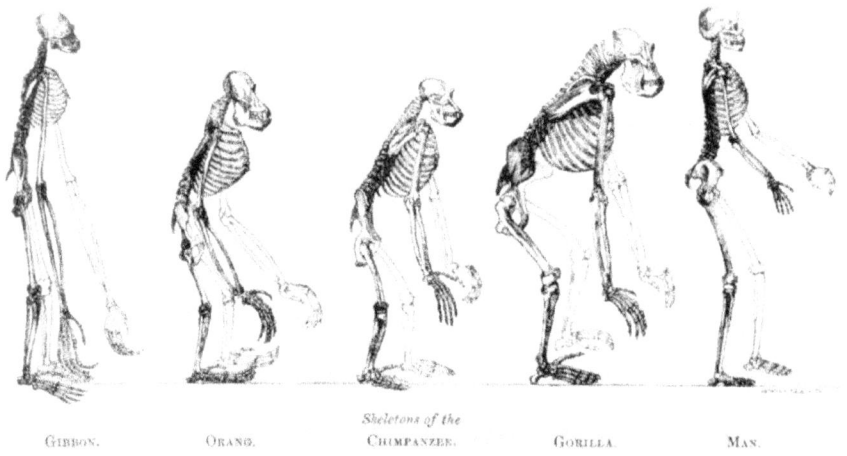

Sheletons of the
GIBBON.   ORANG.   CHIMPANZEE.   GORILLA.   MAN.

Photographically reduced from Diagrams of the natural size (except that of the Gibbon, which was twice as large as nature),
drawn by Mr. Waterhouse Hawkins from specimens in the Museum of the Royal College of Surgeons.

Graph 1

Chimp	Gorilla	Baboon	Lemur	Dog	Chicken	Frog

Questions:

Use Figure 1 on page 257 to compare different ape skeletons to answer the following questions.

1) What do all the skeletons have in common?

2) What do you see is different between the skeletons?

3) How is the human jaw different from the other apes?

4) How is the human braincase different from the other primates?

5) How do you account for these differences in the skull and mandible between humans and the other apes?

6) How are the neck bones different between the apes?

7) Why do you think those neck bones are so different?

8) How do the arms of the human differ from the other apes?

Use Graph 1 on page 259 to answer the questions that follow.

1) Which animal has the most similarities to humans?

2) Which animal has the least similarities to humans?

3) Is that animal in number 2 even a mammal?

4) Which animals are primates?

5) Do primates have more or fewer similarities to humans than non-primates? Explain.

6) Which of the primates have the least similarities?

7) These animals in number 6 are only found in Madagascar. Why do you think this primate is more different than the other primates?

8) Which animals are mammals?

9) Do mammals have more or fewer similarities to humans than non-mammals? Explain why you think this is?

10) Which animal seems to be the most related to the human? Explain Why.

11) Which animal seems to be least related to the human? Explain why.

Hydra Lab

Directions and Questions:

You will need cultures of **hydra** and **Daphnia**, a **dissecting microscope**, a **watch glass**, a **pipette**, a **ruler**, and a **dissecting probe. Looking at the materials and lab we will be using, what are the safety precautions we should take to protect ourselves and materials during the investigation?**

1) Take a pipette and transfer a few hydrae to a watch glass; this animal is related to jellyfish. It is a polyp where we see jellyfish swimming as a medusa stage of its life cycle. Add enough water to cover the hydra. Put the watch glass under the dissecting microscope to focus and observe the hydra. Do not use direct light; use indirect light to observe the hydra.

2) What do you see that tells you the hydra is alive?

3) How many tentacles are on your hydra?

4) Find more hydra; do they all have the same number of tentacles?

5) The **mouth** is in the middle between the **tentacles,** which is the anterior end of the hydra. The **basal disc** is on the other end of the hydra, attaching itself to a surface. The **gastrovascular cavity** is between the mouth and basal disc. Draw a picture of the hydra and label it below.

6) Describe how the hydra moves and changes its shape.

7) Tap on the watch glass; how does the hydra react?

8) Take your probe and touch the hydra on different parts of its body. How does it react when you touch it on different parts of its body?

9) Does just one part of the body react to the touch, or does the whole body react?

10) Take your pipette and add some Daphnia to the watch glass with the hydra. Describe how the hydra captures the Daphnia.

11) What evidence do you see that shows the hydra is stinging the Daphnia with **nematocysts**?

12) How does the hydra eat the food?

13) Hydra is a very simple animal; what are things you do not see that other animals generally have on their bodies?

Observing Flatworms and Roundworms

Directions and Questions:

You will need **prepared slides** of **flatworms** and **roundworms** and a **compound light microscope. Looking at the materials and lab we will be using, what are the safety precautions we should take to protect ourselves and materials during the investigation?**

1) Focus a slide of a flatworm like a tapeworm under the microscope according to your teacher's instructions. Draw a picture of its anatomy here:

2) Focus a slide of a roundworm like a pinworm under the microscope according to your teacher's instructions. Draw a picture of its anatomy here:

3) Which worm looks more complicated?

4) Which worm has a digestive tract?

5) How do you think the tapeworm gets its nutrients?

6) How do you think the roundworm gets its nutrients?

7) Of the worms you looked at today, which is more likely to use movement?

8) Which may be least likely to use movement?

Planarian Lab

Directions and Questions:

You will need a **live culture of planaria** in **pond water**, a **pipette**, a **dissecting probe**, a **Petri dish**, a **scalpel**, and a **dissecting microscope. Looking at the material and lab we will be using, what are the safety precautions we should take to protect ourselves and materials during the investigation?**

1) Pour some pond water into a Petri dish, then use a pipette to take a planarium from the culture and put it in the Petri dish. Place the Petri dish under the dissecting microscope.
2) Observe the planaria under the microscope. How do you see it move?

3) Touch the planaria with the probe in different places. How did it react?

4) Use the chart below to help you get familiar with its anatomy.

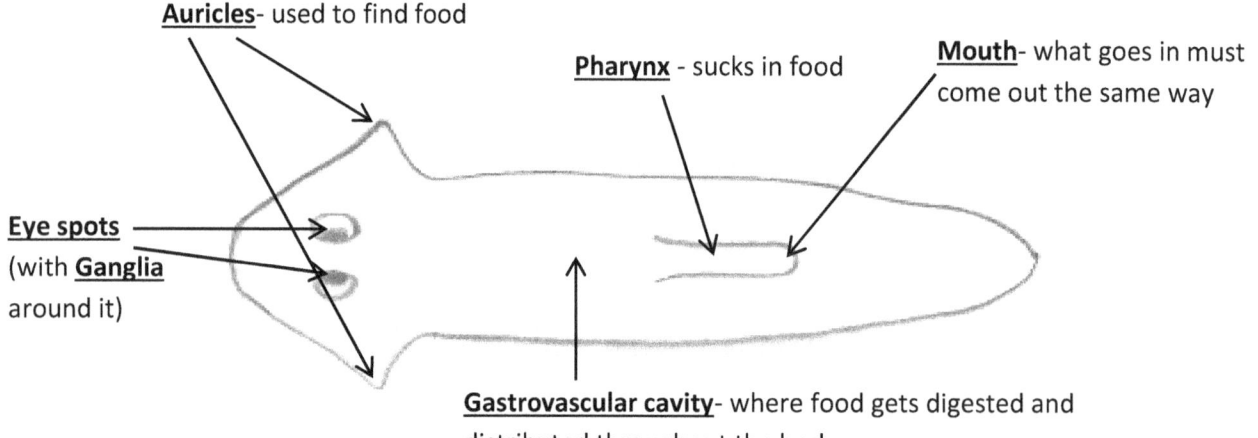

Auricles- used to find food

Pharynx - sucks in food

Mouth- what goes in must come out the same way

Eye spots
(with **Ganglia** around it)

Gastrovascular cavity- where food gets digested and distributed throughout the body

5) Each group will use a scalpel to cut the planaria in different ways, then put the lid on the Petri dish and check the worms later in the week.
   a. Half the groups will just split the head down the middle giving it two heads.
   b. The other groups will make two cuts. They will cut the head and tail off, making three different worms.

6) How did the worms react to being cut?

7) What do you think you will see when you come back and look at them in a few days?

8) After waiting a few days, what happened to the worms you cut up (this is called **regeneration**)?

9) Why do you think it is easy for this animal to regenerate parts but difficult for humans?

Earthworm Dissection

Directions and Questions:

You will need a preserved **earthworm, dissecting scissors, dissecting pins**, a **dissecting pan** (with a rubber bottom), and a **hand lens** or **dissecting microscope. Looking at the materials and lab we will be using, what are the safety precautions we should take to protect ourselves and materials during the investigation?**

1) **External Anatomy**: The **mouth** is located on the front fatter pointed end of the worm just behind the **prostomium** (the first segment of the worm). A saddle-shaped structure called the **clitellum** is closer to this end. On the ventral side of the worm, most of the external structures can be found. Find the worm's ventral side and use the diagram below to help you find the earthworm's external structures.

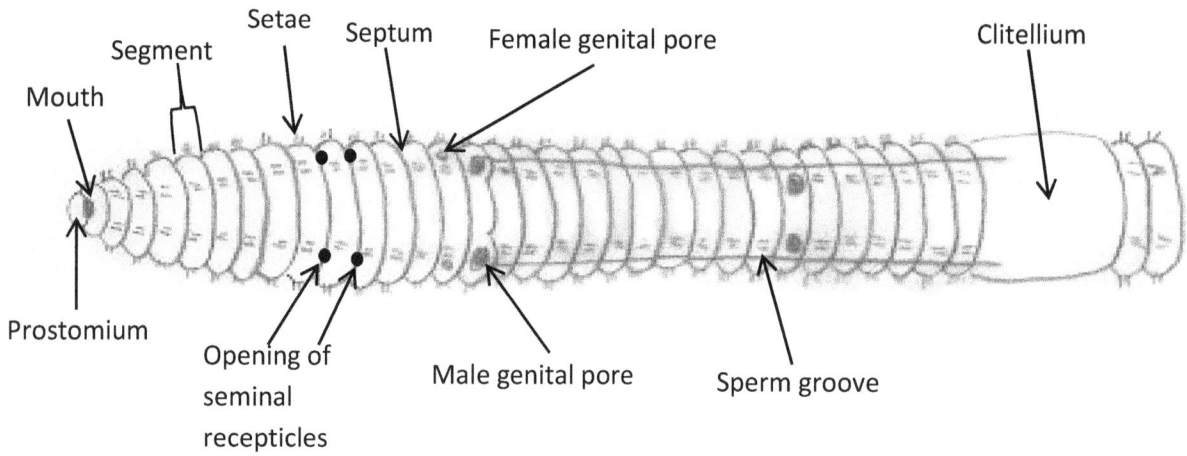

2) There are four **setae** on each segment, two on the left side and two on the right. These give the worm grip underground to move. **Setae** also make it hard for them to be pulled out of the ground.

3) Earthworms are hermaphrodites (they are both male and female at the same time). They will have both sets of sex organs. Starting on the 9th and 10th segments, you will see the openings to the **seminal receptacles**. The **female genital pores** can be found in the 14th segment. The **male genital pores** will be found on the 15th segment and another set on the 26th segment. A sperm groove runs from the front **male genital pores** back to the **clitellum,** used in mating.

4) Past the **clitellum**, different worms have different amounts of segments. Just past the very last segment is the **anus,** where the digested dirt comes out of the worm.

5) **Internal Anatomy**: Take the earthworm and face it so that the dorsal side is up and the ventral side, with the setae down. Take your dissecting scissors and gently cut a hole through the prostomium cutting down the length of the worm. Have the scissor's pointy end in the worm, cutting and lifting up slowly as you go not to damage the worm's internal organs.

6) Gently use your probe and pins to pull back the skin of the worm, pinning it back as you go placing pins on each side of the worm every few segments. Continue to do this back to the clitellum. Now you should have the internal organs exposed for you to see. Use the diagram below to help you identify the internal organs.

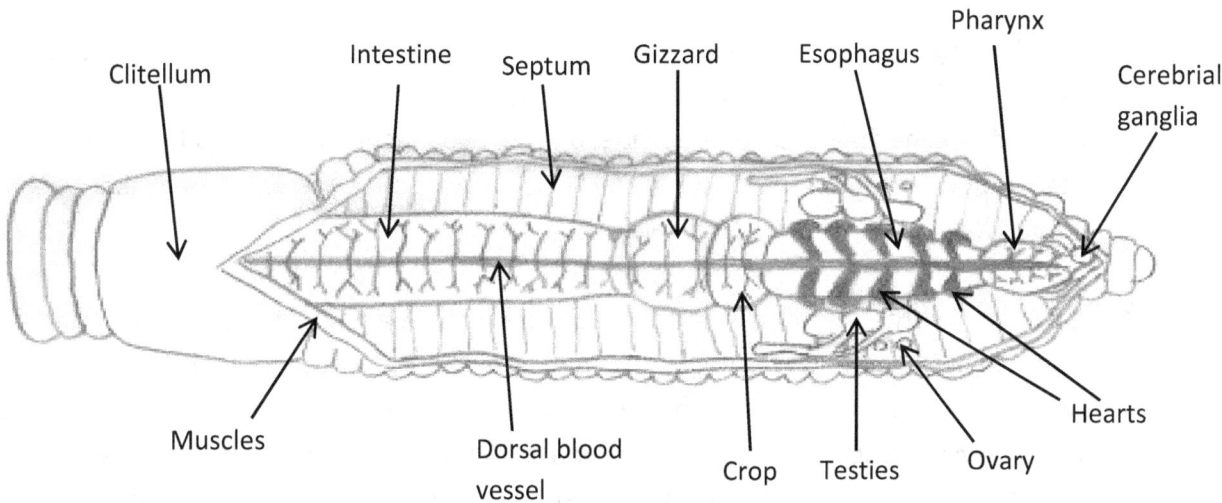

7) The **Digestive tract**: The worm swallows with the muscular **pharynx**. The dirt then passes down the **esophagus** to the **crop,** which is a storage chamber. The **gizzard** is muscular and grinds to break the dirt down more (use your probe to see how different the **crop** and the **gizzard** feel). The dirt is then passed back to the **intestine**, where nutrients are absorbed. Why do you think the intestine is so long, going all the way back to the **anus**?

8) The **Circulatory System**: The earthworm has a **closed circulatory system** (all the blood is contained in blood vessels. There are **five hearts** that surround the esophagus that help pump the blood throughout the worm. You can see the **dorsal blood vessel** as it runs

down the middle of the worm over the digestive tract. A ventral blood vessel is under the digestive tract, also running the worm's length. What do you think are the reasons the dorsal and ventral blood vessels are located there?

9) The **Nervous System**: The two **cerebral ganglia,** which act as a tiny **brain,** sit on the front end of the pharynx. A **ventral nerve cord** can be seen if you gently push the digestive tract aside. It is a cream-colored line that also runs the length of the worm. There will be branches of **nerves** that come off the nerve cord at each segment. Why do you think this is?

10) The **Reproductive System**: Four seminal receptacles look like tiny cream balls, two on each side of the worm just anterior to the larger **testicles**. The **testicles** are large cream-colored organs found on either side of the esophagus. These produce sperm. Just posterior to the **testicles,** you can find the **ovaries** that are small and also cream in color. These produce the worm's eggs.

11) Other structures are the **septa** that separate each segment of the worm. Inside each segment, you can find two **nephridia** that take waste products out of the blood all the way down the worm. One on each side lateral to the digestive tract. There are **longitudinal muscles** that make the worm longer and shorter as it moves. **Circular muscles** make the worm wider or skinnier when they contract and relax. These muscles lay just under the skin and help make up the walls of its body cavity. How do you think the muscles work with the **setae** on the outside to move the worm?

12) Earthworms are simple, but marine worms look more complex. Find a picture of a marine worm and draw it below. After looking at the marine worms, you may be able to see how segmented worms gave rise to arthropods.

Crawfish Dissection

Directions and Questions:

You will need a **preserved crawfish**, **dissecting scissors**, a **dissecting pan** (with a rubber bottom), a **dissecting probe**, and some **dissecting pins**. **Looking at the materials and lab we will be using, what are the safety precautions we should take to protect ourselves and materials during the investigation?**

1) Use the picture below to help you explore the **exoskeleton** of the crawfish.

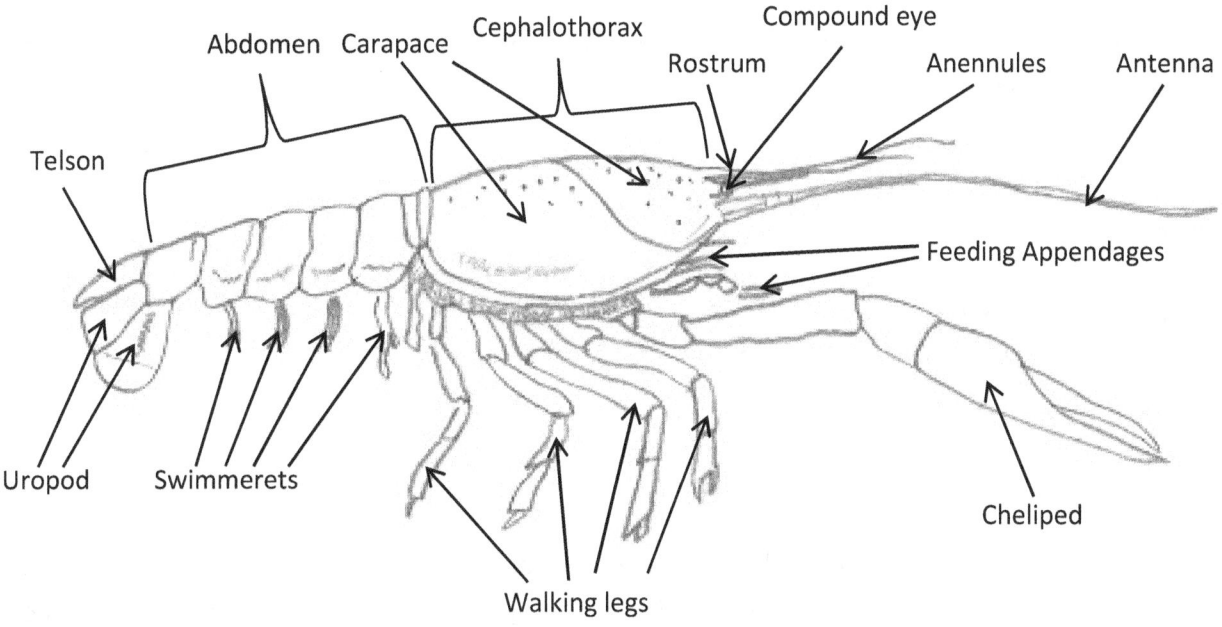

2) At the base of each **antenna** is a **nephridia pore** where the crawfish urinates out of its face. You should be able to fit the end of your probe in the opening. What do you think is the function of the **antenna** and **antennules**?

3) Find the **mouth** on the face of the crawfish. What direction do the **jaws** (**mandibles**) move?

4) Looking at the **cheliped,** how do you think it is used?

5) Looking at the **walking legs,** how do you think they are used?

6) Where do you think the crawfish should be able to taste to detect food (Hint: they touch the ground)?

7) The tail or **abdomen** is very muscular; this is the main part we eat. How do you think the crawfish use it?

8) If you look under the **telson** on the ventral side, you will see a hole; this is the **anus**. How do you think the **telson** and **uropods** are used?

9) Notice that the crawfish has an exoskeleton covering its whole body; this is characteristic of all **arthropods** (which means jointed appendages). What are some other arthropods that you know?

10) Take your dissecting scissors and cut up the thorax from the abdomen to between the eyes. Make sure you pull up on the scissors as you cut so as not to damage the internal organs under the carapace. In front of the eyes, make lateral cuts so that the thorax's exoskeleton can now be opened like doors.

11) Peel back the exoskeleton of the thorax to see the internal organs.

12) Take the dissecting scissors and cut down the abdomen's dorsal side, starting near the thorax moving to the telson. As you cut, pull up on the scissors, so you do not damage the intestine and muscles underneath.

13) Use the diagram below to help you explore the internal organs of the crawfish.

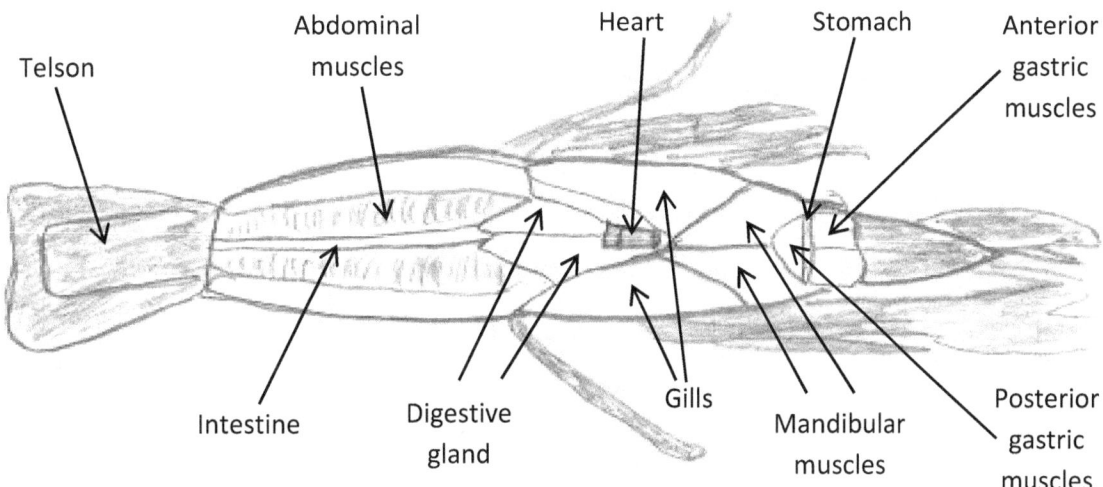

14) If you lift the **stomach** and look on the inside of the face, on each side, you should see **green glands** that filter out metabolic wastes and allow the crawfish to do what out of its face through the **nephridia pores**?

15) Cut into the **stomach** to see that the **teeth** are located there. The **digestive gland** is behind the **stomach** and **gills**. Why do you think the **teeth** are located in the **stomach**?

16) Notice where the **heart** is. The crawfish has an open circulatory system (blood pools throughout the body and is not all contained in blood vessels). How do you think the crawfish uses the **heart**?

17) On either side of the **heart** and **stomach** are the **gills**. How do you think they are used?

18) You have a female if you see bright orange eggs inside on each side. If you do not see the eggs, you have a male. You can also tell the sex with the first pair of **swimmerets**; if they are big, it is a male; if they are small, a female. Do you have a male or female?

19) In the **abdomen**, you can see the **intestine** run down the middle of the **abdominal muscles**. You can turn the crawfish over and see that it empties out the **anus** under the **telson**.

Virtual Investigations that go with Evolution of Animals

ExploreLearning.com

Hardy-Weinberg Equilibrium Gizmo

Human Skull Analysis Gizmo

Embryo Development Gizmo

Unit 10 Animal System Interactions

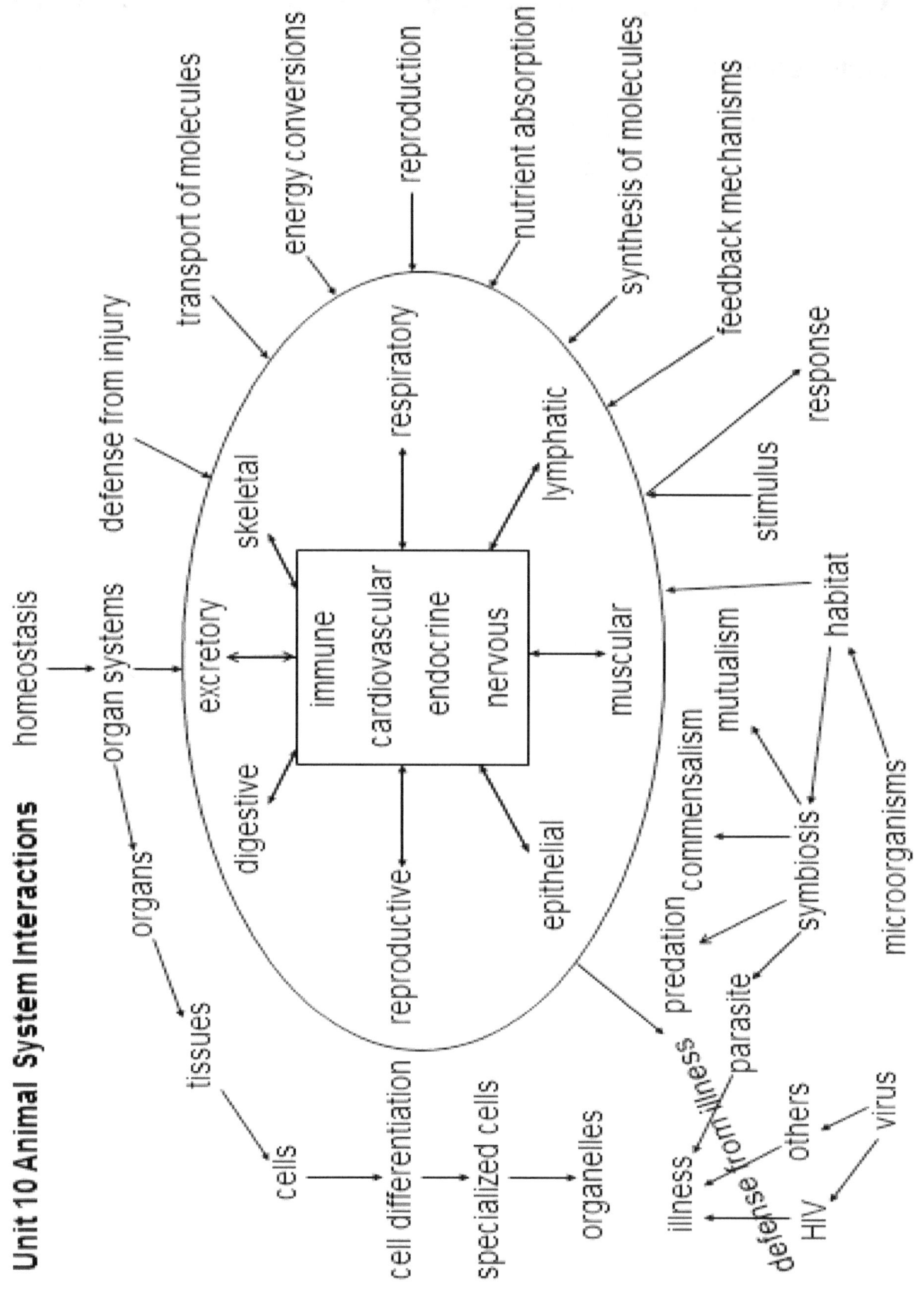

Recognizing Body Tissues

Directions:

You will need **various prepared slides of body tissues** to look at under a **compound light microscope,** and **lens wipes** to be used to clean them off. **Looking at the materials and lab we will be using, what are the safety precautions we should take to protect ourselves and materials during the investigation?**

Focus, look at, and draw the body cells and tissues that are on the slides your teacher has given you in the left-hand boxes below. Describe the cells/tissue features that help you recognize it under a microscope in the right-hand box next to the drawing.

Body Tissue Drawing and Descriptions:

Name of Tissue Drawing:	Description:
Name of Tissue Drawing:	Description:
Name of Tissue Drawing:	Description:

Name of Tissue Drawing:	Description:
Name of Tissue Drawing:	Description:
Name of Tissue Drawing:	Description:
Name of Tissue Drawing:	Description:
Name of Tissue Drawing:	Description:
Name of Tissue Drawing:	Description:

Heart Dissection

Directions:

You will need a **heart model**, a **labeled heart diagram**, **preserved sheep's heart**, a **dissecting pan**, and a **scalpel. Looking at the materials and lab we will be using, what are the safety precautions we should take to protect ourselves and materials during the investigation?**

1) Take your sheep's heart and find the ventral and dorsal sides of the heart. The dorsal side has pulmonary arteries and veins, whereas the ventral side does not. Use your scalpel to cut the heart in half, separating the ventral and dorsal sides (front and back). Ask your teacher for help if you cannot tell which side is dorsal and which is ventral.

2) You should now be able to see all the chambers and valves in the heart. Draw and label the heart in the space below. You can use pictures from your textbook, models you may have in the class, or pictures of a mammalian heart on the internet to help guide you.

3) Draw arrows showing the direction of the blood flowing through the heart.

Drawing:

Questions:

1) Why is there a septum separating the left half of the heart from the right?

2) Why are the walls of the atriums thinner than the ventricles?

3) What are the functions of the valves in the heart?

4) Why are the valves controlled from the chamber where the blood is going?

5) How many valves do you see in the heart?

6) How many chambers does this heart have?

7) Which chamber brings the blood in from the body?

8) What are the blood vessels called that bring in the blood?

9) Which chamber pumps the blood to the lungs?

10) Which blood vessels take the blood from the heart to the lungs?

11) Which chamber of the heart brings the blood in from the lungs?

12) What blood vessels bring the blood from the lungs to the heart?

13) Which chamber pumps the blood to the rest of the body?

14) What is the largest blood vessel in the body that leads out of the heart?

15) Why do you think the left ventricle heart muscle is thicker than the right ventricle?

Making a Working Model Lung

Directions:

You will need a **plastic two-liter bottle**, **strong scissors**, a **round balloon**, **cellophane**, and **packing tape**. **Looking at the materials and lab we will be using, what are the safety precautions we should take to protect ourselves and materials during the investigation?**

1) Take the two-liter bottle and cut the bottom off with scissors.
2) Take the balloon and push it through the small opening in the top of the bottle keeping the balloon's lip on the bottle opening threads and balloon inside the bottle.
3) Take the cellophane and make a diaphragm loosely (so you can move it up and down), covering the bottom of the bottle you just cut off.
4) Take the packing tape and securely fix the cellophane to the bottle, making it airtight.

Questions:

1) What happens to the bottle's volume when you pull the cellophane down?

2) What does this do to the pressure in the bottle when the cellophane is pulled down?

3) When you pull the cellophane down, this is like breathing in. What happens to the balloon when you pull the cellophane down?

   a. Why do you think this happens?

4) What happens to the bottle's volume when you push the cellophane up?

5) What happens to the pressure inside the bottle when the cellophane is pushed up?

6) When you push the cellophane up, it is like breathing out. What happens to the balloon when you push the cellophane up into the bottle?

     a. Why do you think this happens?

7) Which part of the model represented the lungs?

8) Which part of the model represented the chest cavity?

9) Which part of the model represented the diaphragm?

10) How is this model you made similar to how your lungs work in real life?

Measuring Respiration of an Animal

Directions:

You will need a **gas sensor habitat** with two openings, a **live insect or arthropod** (make sure it is a safe one that is not dangerous), and **oxygen gas** and **carbon dioxide gas sensors** attached to an **interface** connected to a **computer** with **Logger Pro. Looking at the materials and lab we will be using, what are the safety precautions we should take to protect ourselves and materials during the investigation?**

1) Make sure the sensors collect data for at least 15 minutes in Logger Pro.
2) Find a place the insect or small arthropod (the bigger, the better) into the habitat and ensure the oxygen and carbon dioxide sensors block the opening so the air is sealed and the insect cannot get out.
3) Press "Collect" in Logger Pro. Press the big "A" in the toolbar to scale the data so it is easier to see.

Questions:

1) What process do all animals have to go through to stay alive?

2) As time went on, what happened to the amount of oxygen in the habitat?

3) What happened to the amount of carbon dioxide in the habitat as time went on?

4) Why do you think the amounts of gases are changing this way?

5) What do you think will happen to our critter in the habitat if we keep it sealed shut for a long time?

Control of Human Respiration

Directions:

You will need a **small bag** to breathe in, a **Vernier respiration monitor belt** connected to a **Vernier gas pressure sensor** attached to an **interface** connected to a **computer** with **Logger Pro**. **Looking at the materials and lab we will be using, what are the safety precautions we should take to protect ourselves and materials during the investigation?**

1) Open Biology with Vernier and then file #26a Human Respiration in the Logger Pro file folder.

2) Wrap the respiration belt around one of the student's chest. Have them sit with their back to the computer monitor so they cannot see the data. Close the valve on the bulb pump by turning it clockwise as far as it goes. Squeeze the bulb to pump air into the belt as much as possible without the student feeling uncomfortable. Have the student breathe and see if you can see the pressure rise when they breathe in and fall when they breathe out (see a difference of at least 2-3 kPa). If not, then pump more air into the belt.

3) Have the student breathe normally. Start collecting data by clicking the "Collect" button. After 60 seconds go by, have the student hold their breath for 30 to 45 seconds. Once they release their breath, have the student breathe normally for the rest of the time.

4) Determine the respiration rate before and after the student held their breath. The examine button may help you determine these values. Put this data in Data Table 1.

5) Go back to the Logger Pro file folder and open file #26b Human Respiration.

6) Get a different person to test this time. Repeat the procedure in #2 to get them set up.

7) Have this student breathe into a small bag during the data collection. If this person gets dizzy, nauseous, or a headache, stop the experiment immediately. Click "Collect" and collect the data for 300 seconds.

8) Once the data is collected, find the maximum and minimum pressure for the first 30 seconds, seconds 120-150, and seconds 240-270. Do this by highlighting the graph section and clicking the "STAT" button. Subtract the minimum and maximum pressure in kPa. Record these values in Data Table 2.

Data Table 1

Before holding breath	After holding breath
breaths/minute	breaths/minute

Data Table 2

0-30 seconds	120-150 seconds	240-270 seconds
Maximum: kPa	Maximum: kPa	Maximum: kPa
Minimum: kPa	Minimum: kPa	Minimum: kPa
Difference: kPa	Difference: kPa	Difference: kPa

Questions:

1) Use Data Table 1: Did the respiratory rate change after holding their breath? If so, how did it change?

2) Use the graph on the computer: What happened to the amplitude and frequency of the graph as time went on that helped make Data Table 2?

3) Why do you think this happened?

4) How do you think carbon dioxide affects your breathing?

Human Respiration

Directions:

You will need a **round balloon**, a **ruler**, a **meter stick**, a **bread bag**, a **rubber band**, and an **oxygen sensor** attached to an **interface** connected to a **computer** with **Logger Pro**. **Looking at the materials and lab we will be using, what are the safety precautions we should take to protect ourselves and materials during the investigation?**

Part 1

1) Before starting, make sure you stretch out the balloon. Take a normal breath and exhale normally into the round balloon (do not force air into the balloon). Pinch the balloon so that no air escapes. Use the metric side of a ruler to measure the diameter of the balloon. Record this measurement in Data Table 1.

2) Repeat the procedure in #1 two more times. Take your three measurements and find the average by adding them up and dividing them by three. Record this in Data Table 1.

3) Now repeat the process in #1, except inhale as much as you can and exhale as much air as you can into the balloon. Repeat the process two more times so that you have three trials. Record this data in Data Table 1 for Vital Capacity.

4) Use Graph 1 to convert the two average balloon diameters in Data Table 1 into lung volumes. Use the balloon's diameter on the "X" axis to show where to look on the graph's curve to give you the point and "Y" value. This "Y" value is the lung volume in cubic centimeters (cc or cm³). Write this information at the bottom of Data Table 1.

Data Table 1

Trial	Normal Breath (Tidal Volume)	Vital Capacity
1	cm	cm
2	cm	cm
3	cm	cm
Average	cm	cm
Lung Volume	cm³	cm³

Graph 1

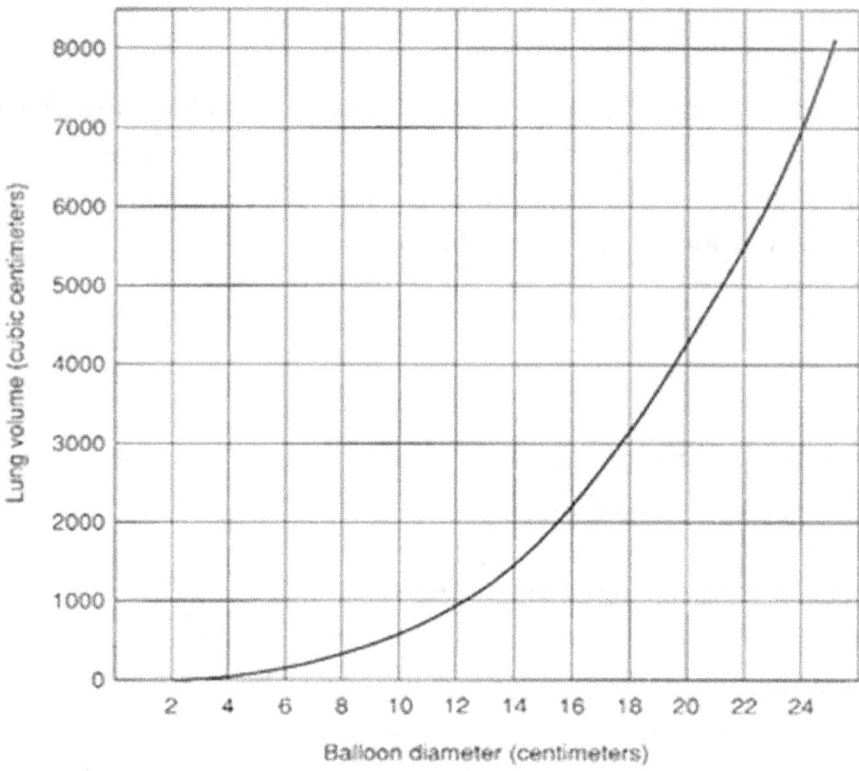

Part 2

5) Make sure your Oxygen sensor is properly connected to the interface and computer. You should see a green button at the top of the Logger Pro screen. Go to the folder in Logger Pro and open Biology with Vernier, and then file #30 Oxygen and Human Reparation. Fix the bottom of the bread bag to the opening of the oxygen probe. Do this by poking a small hole into the bag's bottom and stretching it over the oxygen probe opening. Adding a rubber band should seal the bag to the probe.

6) **Normal Breath**: Have someone take a deep breath and hold it. Immediately click the "Collect" button on the computer in Logger Pro; this will start the data collection. Have the person hold their breath as long as possible. When the person cannot hold their breath anymore, they should blow their breath into the bread bag. Twist the opening of the bag shut immediately, sealing the air inside. Let the data collection finish the 120 seconds.

7) When the data collection is finished, open the bag and pull the opening back past the oxygen sensor to expose the sensor to the fresh air for the next data collection.

8) Click on the Statistics button on the Logger Pro to show the maximum and minimum oxygen levels. You can also find how long the person held their breath by seeing the time when the oxygen level started to drop. Write this data in Data Table 2.

9) Calculate the oxygen change by subtracting the minimum value from the maximum value. Write this data in Data Table 2.

10) **Hyperventilation**: Pull the opening of the bread bag back to what it was before. Have the person now take ten quick deep breaths, then take a large 11th breath and hold it as long as they can. Immediately click the "Collect" button. When they cannot hold it anymore, have them blow their breath into the bag and twist it shut like before, sealing the air inside. Let the data collection finish the 120 seconds.

11) On the Logger Pro, click the statistics button to get the maximum and minimum oxygen levels. You can also find how long the person held their breath by looking at the graph and seeing when the oxygen level begins to drop. Write this data in Data Table 2.

12) Calculate the oxygen change by subtracting the minimum value from the maximum value. Write this data in Data Table 2.

Data Table 2

Type of Breath	How long breath was held	Maximum O_2	Minimum O_2	Change in O_2
Normal				
Hyperventilation				

Questions:

1) Are your tidal volume and vital capacity the same as everyone else in the class?

2) Why do you think that is?

3) Why might it be important to know someone's tidal volume and vital capacity?

4) Why is it that there is still some air left in your lungs when you exhale as hard as you can?

5) What do you think will happen to your vital capacity if you start training for a sport?

6) As you get older and sit around more, what do you think will happen to your vital capacity?

7) How might vital capacity be important to musicians?

8) How do you think smoking and vaping would affect your vital capacity?

9) When you hold your breath, what happens to the oxygen concentration? Why did this happen?

10) What do you think happens to your carbon dioxide concentration when you hold your breath? Explain why.

11) How do you think smoking and vaping would affect the change in oxygen?

12) On average, people can hold their breath for a minute. What do you think prevents people from holding their breath for two to three minutes?

13) How did the amount of time you hold your breath change after hyperventilation? Why did this happen?

14) Was the concentration of oxygen in your breath higher or lower when you hyperventilated? Explain why.

15) When you hold your breath, what do you think forces you to start breathing again?

16) What could be some sources of error in this investigation?

Measuring Heart Rate and Physical Fitness

Directions and Observations:

Hook up the **hand-grip heart rate monitor** to an **interface** connected to a **computer** with **Logger Pro**. Go to the manila folder in the upper left-hand corner of the toolbar in Logger Pro and open Biology with Vernier and file #27 Heart Rate and Physical Fitness. **Looking at the materials and lab we will be using, what are the safety precautions we should take to protect ourselves and materials during the investigation?**

1) Grab the Hand-Grip Heart Rate Monitor in both hands and sit in a chair. Do not let go of the monitor during the entire experiment. After about one minute, press collect and stay seated in the chair for 1 minute. What was the heart rate where the person leveled out? **This heart rate is their resting heart rate.**

2) Once one minute was up, stand up, and for 1 minute. What was the HR at which the person leveled out while standing?

3) After that minute is up, have the person repeatedly squat for 1 minute; what was the maximum HR?

4) Rest for 2 minutes in the chair. Find the slope of the line declining (later after the experiment is over).

5) Repeat steps 2 and 3. What was the maximum HR while exercising?

6) Then rest while standing for 2 minutes. Find the slope of the line declining after the experiment is over.

Questions:

1) When was the heart rate the lowest?

2) When was the heart rate the highest?

3) What caused the HR to go up?

4) What did the body's muscles need more of when the HR was at its max?

5) What did the body need to have taken out of it?

6) How was the body able to do this?

7) Compare the rate at which the HR declined for steps 4 and 6. Which looks like the quickest way to recover from exercise?

8) How do you think the person's breathing changed during the experiment?

9) Why was it changing?

Observing Vertebrate Skeletons

Directions:

You will need **models** of different **animal skeletons** (Halloween decorations sometimes work well) and actual **bones** from animals. **Looking at the materials and lab we will be using, what are the safety precautions we should take to protect ourselves and materials during the investigation?**

1) Take the bones from page 295, cut them out, and tape them together. Use the labels on this skeleton to label the following skeleton's homologous bones. Make sure you use all the bones labeled on the human as labels for each of the three animals that follow. Use labeling strategies shown to you by your teacher.

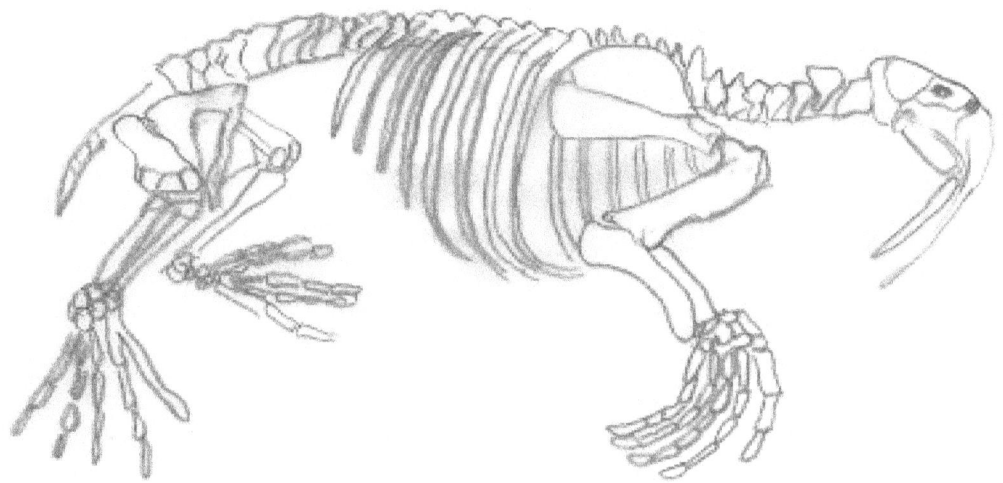

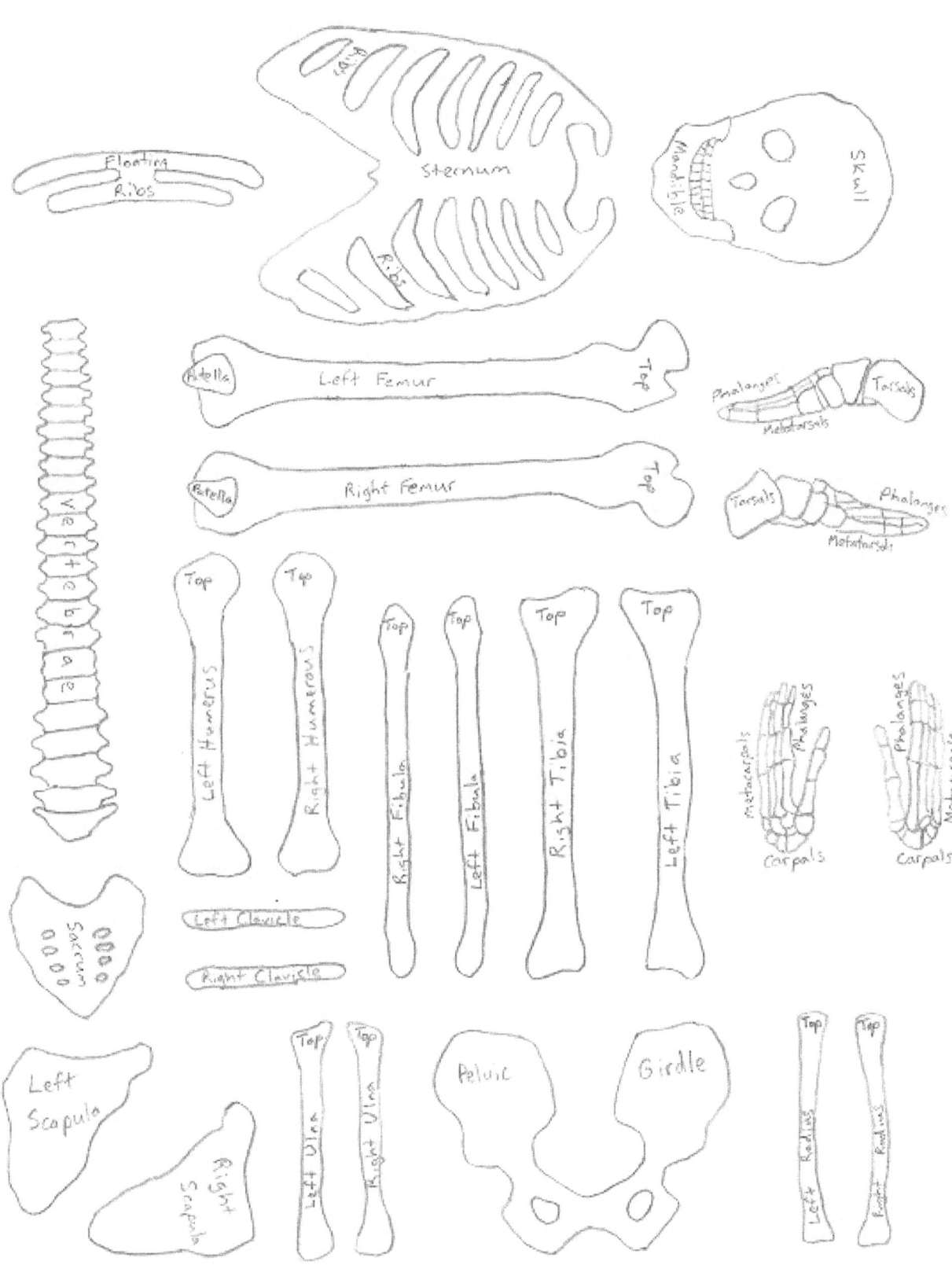

This page will be cut from the previous page.

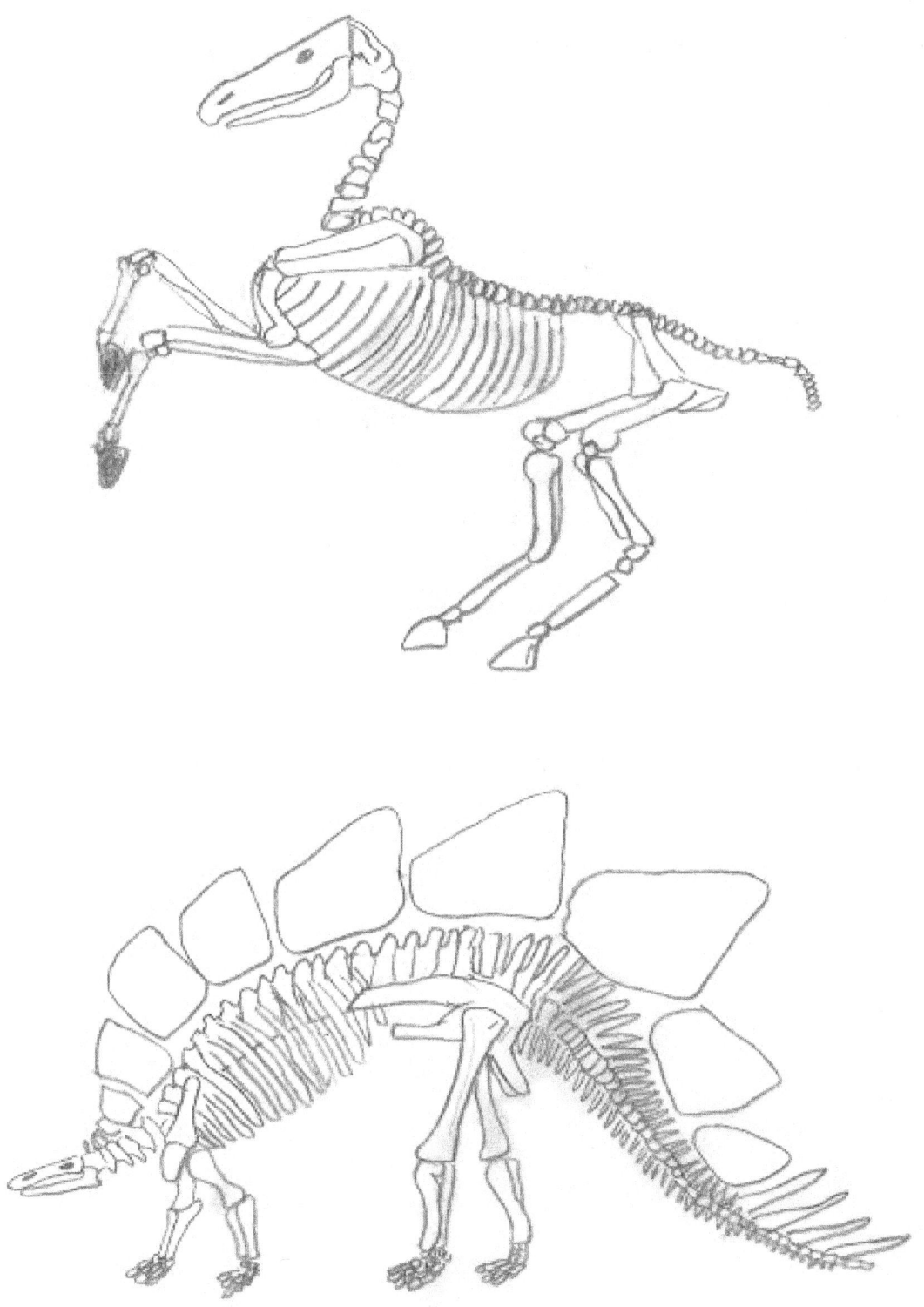

Questions:

1) What did all these skeletons have in common?

2) What was different about each of these skeletons?

3) How were the skulls similar? How were they different?

4) How did the feet differ in each of the skeletons?

5) Look at the bone your teacher gave you; which bone do you think it is?

6) What kind of animal do you think had that bone?

7) What are the functions of the forelegs and hind legs of each of the skeletons?

8) What evidence do you see that all of these animals have a common ancestor with each other?

9) Which one looks most closely related to humans? Explain why.

Brain Dissection

Directions and Questions:

You will need a **labeled diagram** of a **mammalian brain**, a **preserved sheep brain**, a **scalpel**, a **dissecting tray,** and the **internet** or a **textbook. Looking at the materials and lab we will be using, what are the safety precautions we should take to protect ourselves and materials during the investigation?**

1) Draw a lateral view of the brain and label: Cerebrum, cerebellum, pons, spinal cord, medulla, optic chiasma, pituitary, and olfactory bulb.

2) Draw an inferior view and label: optic chiasma, hypothalamus, cerebral peduncles, pons, pyramids, medulla, olfactory bulb, cerebrum, rhinencephalon, spinal cord, midbrain, and cerebrum.

3) Cut the brain in half, separating the left and right sides. Draw a cross-section and label: the Corpus callosum, thalamus, hypothalamus, pons, medulla, arborvitae, and midbrain.

4) Use your textbook and the internet to find the function of each part of the brain that you drew and labeled:
   a. Cerebrum

   b. Cerebellum

   c. Pons

   d. Spinal cord

   e. Medulla

   f. Optic chiasma

   g. Pituitary

   h. Olfactory bulb

   i. Corpus callosum

   j. Thalamus

   k. Hypothalamus

    l. Arborvitae

    m. Midbrain

    n. Cerebral peduncles

    o. Pyramids

    p. Rhinencephalon

5) Look at the survey of brains below; this shows us the evolution of the vertebrate brain. Use this picture to describe how the brains have evolved from fish to mammals. Write the description under the pictures at the bottom of the page.

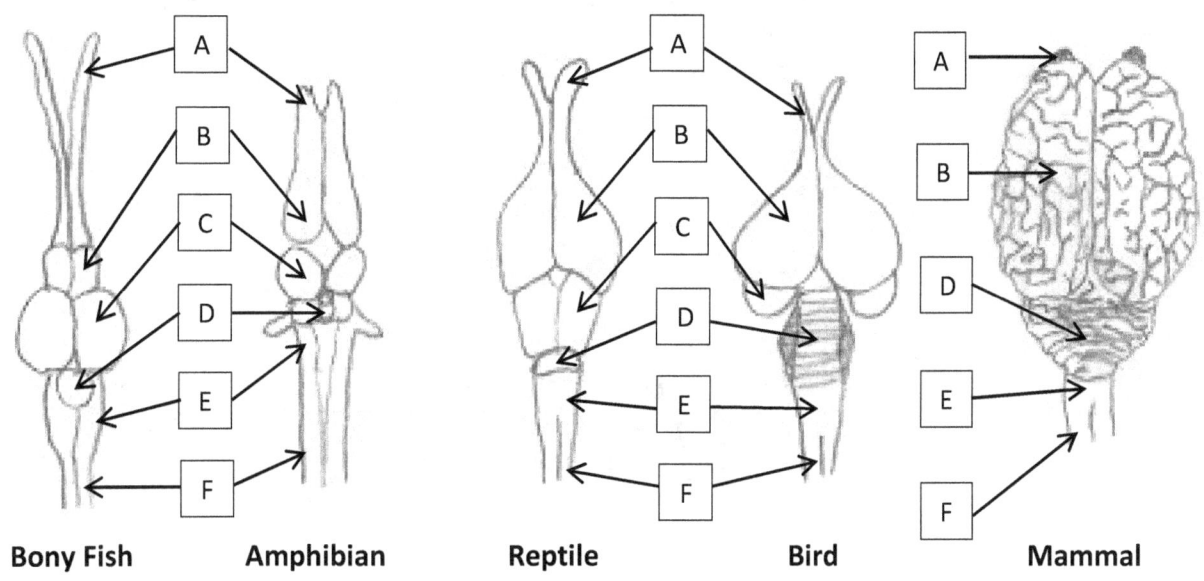

Bony Fish   **Amphibian**   **Reptile**   **Bird**   **Mammal**

A) Olfactory lobe   B) Cerebrum   C) Optic lobe   D) Cerebellum   E) Medulla   F) Spinal cord

Eye Dissection

Directions and Questions:

You will need a preserved **cow eye**, a **dissecting tray**, a **scalpel**, a **dissecting probe**, and **dissecting scissors**. Use the diagram of the eye below to help you with the investigation. **Looking at the materials and lab we will be using, what are the safety precautions we should take to protect ourselves and materials during the investigation?**

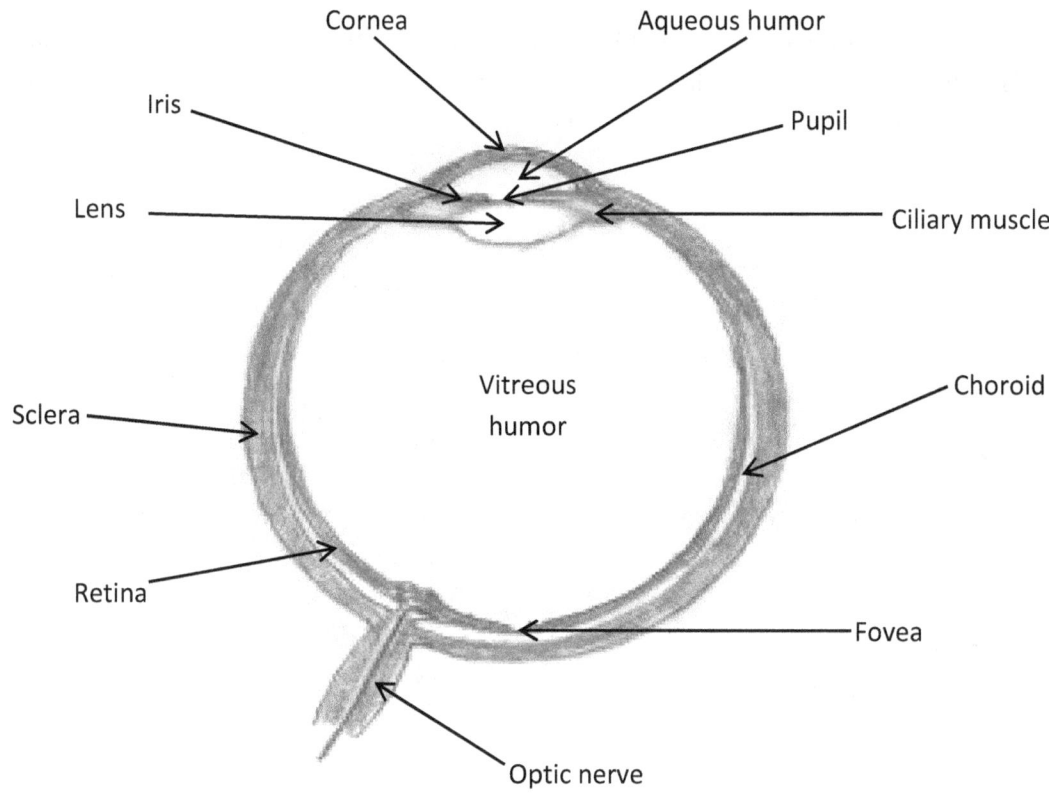

1) Look at the outer surface of the eye. Find the **optic nerve** and the **muscles** surrounding the side and back of the eye. How do you think these muscles were used?

2) Find the **pupil** and **cornea**. What is the shape of the **pupil**?

3) You should hold the eye firmly in your hand with the index finger and thumb (over your dissecting pan). Use a scalpel to make an incision into the **sclera** (the white of the eye), a quarter of an inch away from the **cornea**. Then insert the scissors into the incision and cut all the way around the cornea (be careful not to squeeze the fluid out of the eye).

4) Carefully lift the front part of the eye off the rest of the eye and place it in the tray. Make sure you keep the opening of the eye facing upward.

5) Using the dissecting probe, separate the **lens** from the **vitreous humor** by sliding the probe around the **lens**. Hold the **lens** up and look through it; what do you notice?

6) Use your probe to compare the consistency of the center of the **lens** with the edge. What do you notice?

7) Now look back at the inner surface of the front of the eye you cut off and placed in the tray. Find the thickened black **ciliary muscle**. What do you think is its function?

8) Examine the **iris**. What do you think are the functions of the **circular** and **radial muscle fibers**?

9) If there is any **aqueous humor** left between the **cornea** and **iris**, compare its consistency with the **vitreous humor**. How are they different?

10) Looking back at the back portion of the eye, find the **retina**. It is the thin inner surface of the back of the eye. It can be easily separated from the **choroid**; this shows how easy it is for people to get a detached retina. How do you think a detached retina would change your vision?

11) Find the **blind spot**; it is where the **retina** is attached to the back of the eye. Look behind the eye and see what is right there. What is there that we saw before?

12) Why do you think the **retina** is stuck there and there is a **blind spot**?

13) Notice the choroid's reflective surface (called the **tapetum lucidum**); this causes animals' eyes to reflect light at night in the back of the **retina**. How do you think this will help the animal?

14) Find out what happens at the **fovea**.

Kidney Dissection

Directions and Questions:

You will need **kidney diagrams**, a preserved **sheep kidney**, a **dissection pan**, and a **scalpel**. **Looking at the materials and lab we will be using, what are the safety precautions we could take to protect ourselves and materials during the investigation?**

1) Find a labeled picture of a kidney either in your textbook or on the internet. Obtain your kidney, draw it, and label the **cortex**, **renal hilum**, **renal artery**, **renal vein**, and **ureter** below.

2) Take the scalpel and cut the kidney in half like a hamburger bun so you can see a cross-section of the whole kidney. Draw and label the **renal cortex**, **renal medulla**, **calyx**, **renal pelvis**, **renal artery**, **renal vein**, and **ureter** below.

3) Look at the diagram below and use your textbook or internet to determine the functions of the nephron's labeled parts to see how the kidneys function. Write the function next to the labeled term.

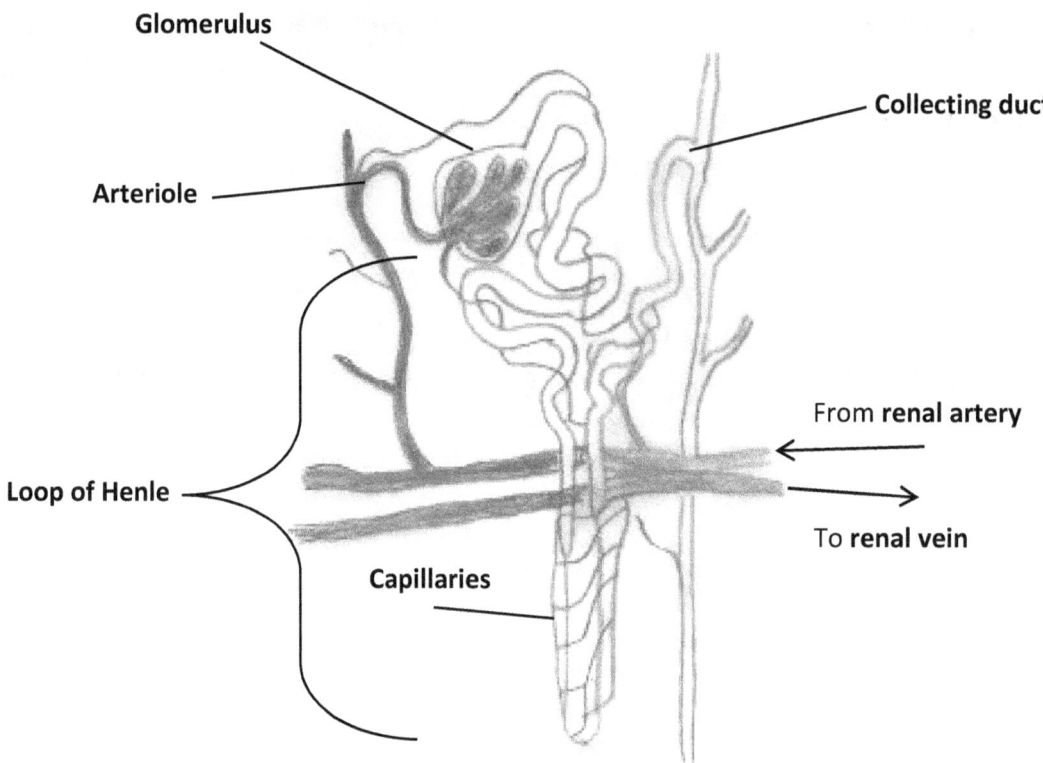

4) Why can't we live without our kidneys?

Ain't Nothing but a Chicken Wing

Directions and Observations:

You will need a **full chicken wing** and a **paper towel**. If it is ok with your teacher, it can be fried so you can eat it when the lab is done. You will also need **prepared slides** of **skeletal muscle**, **smooth muscle**, and **cardiac muscle** that you will look at under a **compound light microscope**. **Looking at the materials and lab we will be using, what are the safety precautions we should take to protect ourselves and materials during the investigation?**

1) If you need to, remove the batter and skin. It should be easy to peel off if cooked because the skin's protein has been denatured by cooking it.

2) The meat is the **muscle**. Notice the **muscle** is attached to the **bone** with **tendons**. One **tendon** is the **origin,** and the other is the **insertion**. The **muscle** does not move the **bone** at the **origin,** but it moves the **bone** at the **insertion** when contracted. Draw a picture of the chicken wing below, labeling: **muscle (bicep and triceps), bone, tendon, ligament, and cartilage.**

3) If you remove/eat the **muscle** off of the **bone**, what do you notice about the **bone** where the **muscle** was inserted (we call it a **tuberosity**)?

4) Notice at the bones' end is **cartilage;** they are smoother and softer than **bone** but harder than **muscle** and not fun to eat. What do you think their purpose is in the joint?

5) Attaching the bones together are **ligaments**. Why do you think they do this?

6) The way the muscles contract is explained through the **sliding filament theory**. The pictures below show the **actin** and **myosin filaments** getting pulled across each other by the **cross-bridges** at **ATP sites**. The **cross-bridges** attach outside and pull inside, shortening the muscle fiber. **Cross bridges** cannot attach and pull without the energy of **ATP**. How do you think your muscle gets stronger?

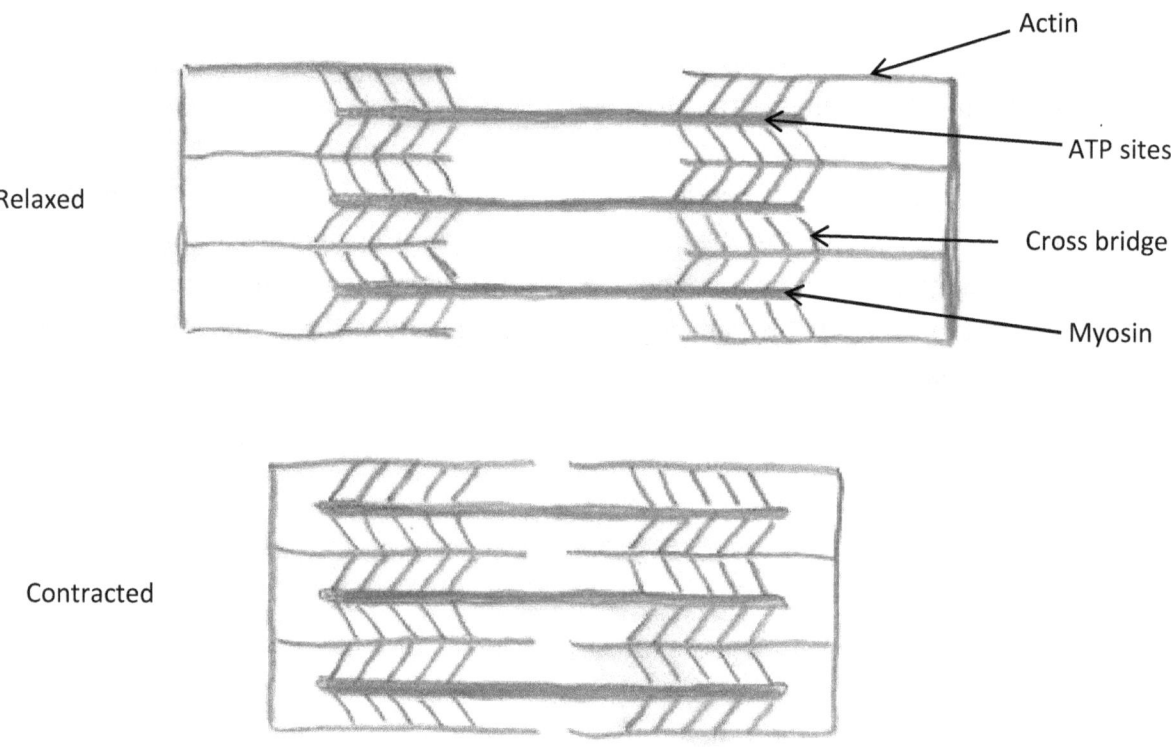

Relaxed

Contracted

7) There are different types of muscle tissue your body uses.

   a. Skeletal muscle is what you are looking at in the chicken. What is its function?

   b. Smooth muscle is used in arteries and the digestive tract. What do you think is its function?

c. Cardiac muscle is found only in the heart. What is its function?

      i. What if the cardiac muscle ever got too tired? What do you think would happen?

8) Obtain a slide of each type of muscle and focus the slide according to your teacher's instructions. Draw pictures of each below.

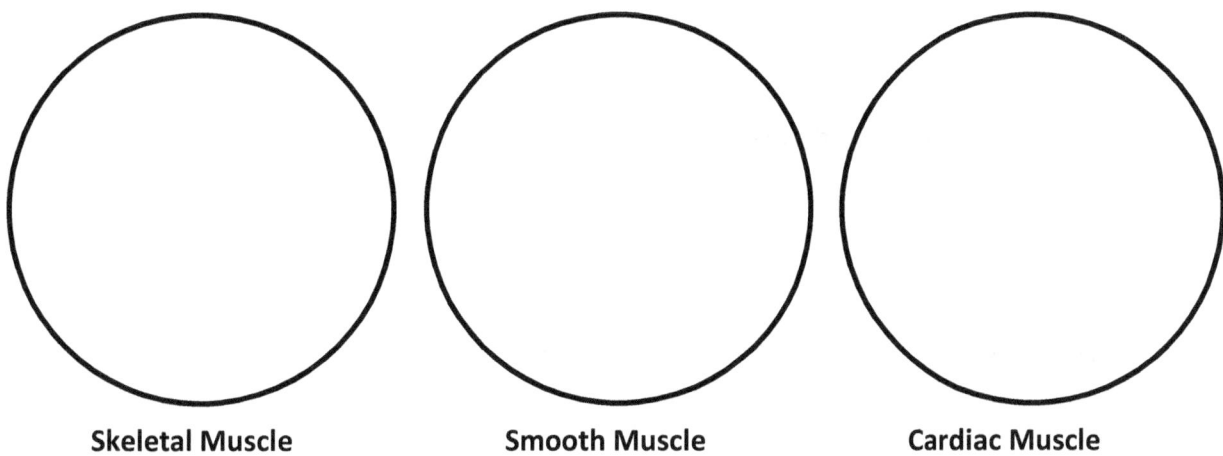

Skeletal Muscle   **Smooth Muscle**   **Cardiac Muscle**

9) Fill in Data Table 1 on what you know and observed from each muscle on the slides.

Data Table 1

Muscle Type	Number of nuclei per cell	Striations (yes/no)	Cell Shape	Voluntary or involuntary
Skeletal				
Smooth				
Cardiac				

Questions:

1) What allows the chicken to move its wing?

2) What would happen if a tendon tore off the bone?

3) How do bones help with movement?

4) How are skeletal and cardiac muscles similar?

      a. How are they different?

5) How are cardiac and smooth muscles similar?

      a. How are they different?

6) What could happen if one of the two opposing skeletal muscles was much stronger than the other?

7) What other body system is used to control all the actions studied in this lab, but we did not talk about it?

Fish Dissection

Directions and Questions:

You will need **safety goggles**, a **preserved perch**, **dissecting scissors**, **forceps**, a **dissecting microscope** or **hand lens**, and a **dissecting pan. Looking at the materials and lab we will be using, what are the safety precautions we should take to protect ourselves and materials during the investigation?**

1) **External Anatomy:** Take your fish and place it in the dissecting pan. Use the diagram below to help you see all its external parts.

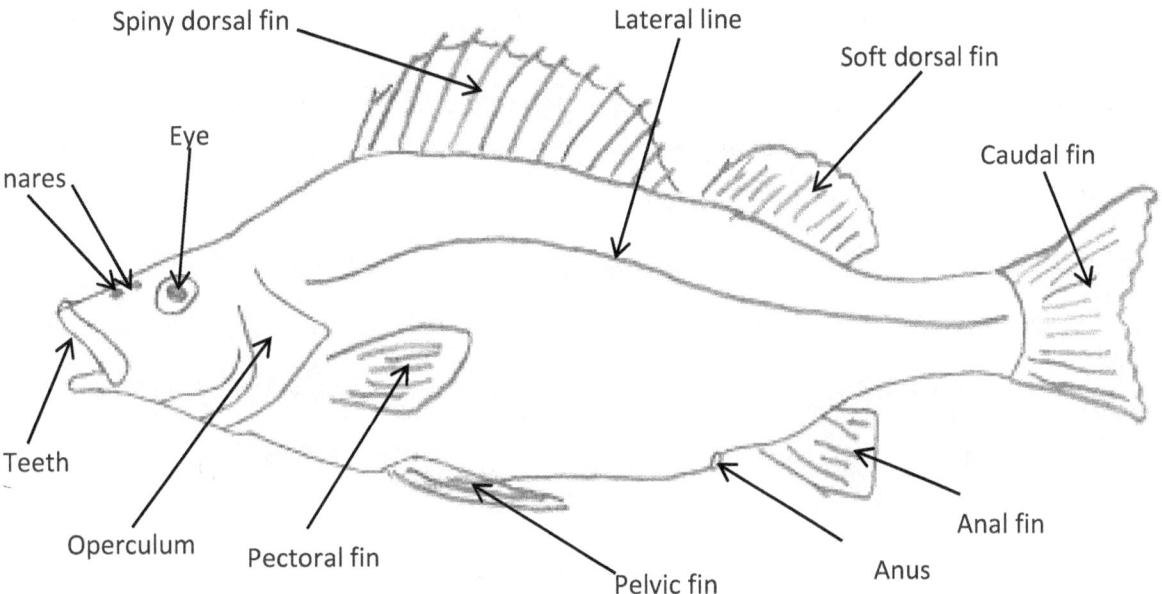

2) The dorsal and anal fins help keep the fish swimming upright in the water. How do you think the spiny rays on the dorsal fin are used?

3) Notice the **lateral line**. It is used to detect motion and vibrations in the water. How could this be useful to the fish?

4) Find the **operculum**. Lift it and see the **gills** underneath. Carefully cut off the **operculum** with the dissecting scissors. What do you think is the function of the operculum?

5) Pull a **scale** off of the fish. Look at it under a dissecting microscope or magnifying glass. What do you see on the scale?

6) As a fish grows, it does not get new **scales**. The **scales** get bigger. What do you think the rings on the **scale** tell you (like a tree)?

7) **Internal Anatomy**: Take your fish and use the dissecting scissors to cut a slit along its belly from the **anus** to the **jaws**. Pull the scissors away from the fish as you do this to not cut through the internal organs. Now make two cuts at each end of the last cut up the fish's side, making a door/flap. Cut up until you see the clear cellophane-looking **swim bladder**. Try not to puncture the swim bladder as you make your last cut, cutting off the flap with the **muscle** and **skin**. Use the picture below to help you find the internal structures of the fish.

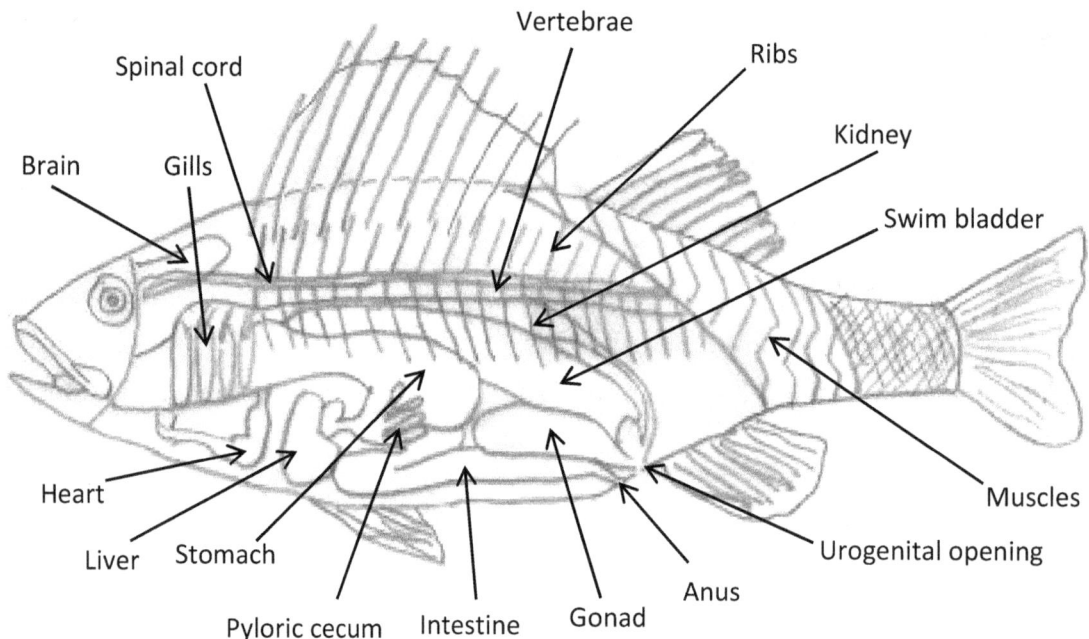

8) Notice the **heart** is located just under the **gills**. Why do you think they are so close to each other?

9) The **liver** is used to remove poisons from the fish's blood. Why do you think it is so close to the **stomach** and **heart**?

10) The **gonad** will either be a sack of **eggs** or a pale **testicle**. If you see an orange sac of **eggs**, you have a female; if you see the cream-colored **testicle,** you have a male. What is the sex of your fish?

11) See how long the **digestive tract** is from the **throat** to the **anus**. This **digestive tract** is not very long compared to mammals. Why do you think fish have a shorter **intestine** than mammals?

12) You may have seen some lobes or flaps coming off the **stomach**. These are the **pyloric cecum**. These will store hard-to-digest food. How is this useful to the fish?

13) Find the organ that looks like a bubble above the **digestive tract**. This organ is called the **swim bladder**; this holds air to make the fish more or less buoyant in the water. The **swim bladder** evolved from a primitive lung from its ancestors but is not used as one in the perch. What do you think the **swim bladder** will help a fish do?

14) Above the **swim bladder** is a flat pinkish-looking **kidney** that filters out the fish's metabolic waste. Trace it and see; where does it exit out of the fish?

15) Above the **kidney** is the **vertebra** and **ribs** that protect the **spinal cord** and are used for muscle attachment sites. Why do you think the rest of the fish is **muscle** (which is most of the fish)?

Dissection Comparing a Rat to a Human

Directions:

You will need **safety goggles**, a **preserved rat**, **dissecting scissors**, a **dissecting pan** with a rubber bottom, some **string**, **dissecting pins**, a **human torso model**, and **labeled diagram(s) of internal organs**. Looking at the materials and lab we will be using, what are the safety precautions we should take to protect ourselves and materials during the investigation?

1) Take your dissecting scissors and cut a slit up the rat's belly, starting in the groin area up to the lower jaw. Try just to cut the layer of skin, do not go into the muscle. To skin the rat, carefully cut slits down each leg. Then carefully pull the skin back, working your way around the rat.

2) Once you have the skin off, notice the muscle structures that are around the rat. They are very similar to ours. They have the same muscles; they are attached in slightly different positions because of the skeletal structure. Locate each of the muscles using the human muscle diagram below. Put a checkmark by the label of muscles below that you also found on the rat.

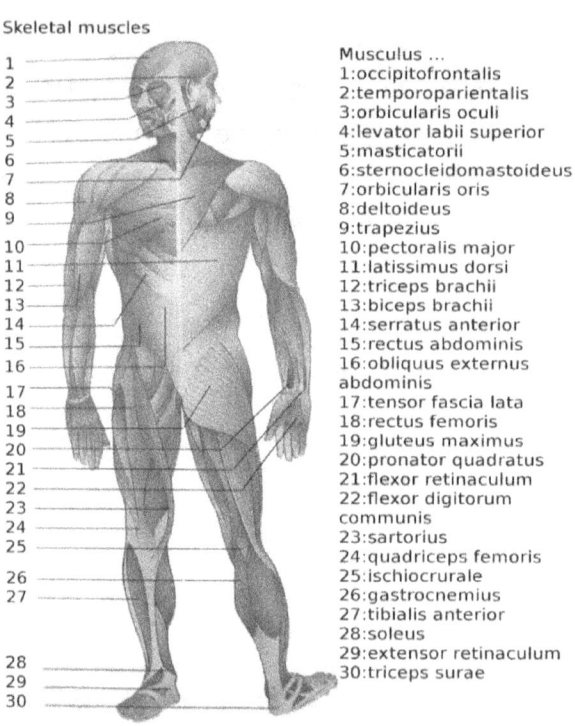

Skeletal muscles

Musculus ...
1:occipitofrontalis
2:temporoparientalis
3:orbicularis oculi
4:levator labii superior
5:masticatorii
6:sternocleidomastoideus
7:orbicularis oris
8:deltoideus
9:trapezius
10:pectoralis major
11:latissimus dorsi
12:triceps brachii
13:biceps brachii
14:serratus anterior
15:rectus abdominis
16:obliquus externus abdominis
17:tensor fascia lata
18:rectus femoris
19:gluteus maximus
20:pronator quadratus
21:flexor retinaculum
22:flexor digitorum communis
23:sartorius
24:quadriceps femoris
25:ischiocrurale
26:gastrocnemius
27:tibialis anterior
28:soleus
29:extensor retinaculum
30:triceps surae

Retrieved From Commons.wikimedia.org

3) Take your rat and put it into the dissecting pan. Tightly tie one end of the string around one wrist of the rat. Take the string, pull it behind the dissecting pan, spread the rat's arms as wide as you can, and tie the other wrist holding the arms wide open. Do the same with the hind legs.

4) Now take your dissecting scissors and cut a slit up the rat's belly through the muscle, starting in the groin area up to the lower jaw. Make sure to pull up as you do so as not to cut into and damage the internal organs underneath. It will get hard as you go through the ribs, but the scissors should be able to cut through.

5) Then we will create two doors/flaps by cutting from the bottom of your slit around each of the hind legs to the back. Then do the same up by the shoulders cutting around the neck. You may also have to cut the abdominal muscles free from the diaphragm above the stomach and below the lungs. You should now be able to pull these flaps back and pin them down.

6) Now you have the internal organs exposed. These are in the same places as humans. Use the human torso in your class and the labeled diagram that goes with it to help you find all your rat's internal organs. Then use the chart below to check off similarities between the two.

Data Table 1

Organ	Found in rat	Found in the same place
Trachea		
Esophagus		
Heart		
Lungs		
Diaphragm		
Liver		
Gall bladder		
Stomach		
Pancreas		
Small intestine		
Large intestine		
Spleen		
Kidneys		
Ureter		
Urinary bladder		
Urethra		
Brain		
Eyes		

Questions:

1) Which muscle(s) could you not find on the rat that was labeled on the human?

2) Which organs on your human torso could you not find on your rat?

3) Which internal organs of the rat were not in the same place as the human?

4) How is the rat's anatomy similar to the humans?

5) How is the rat's anatomy different from humans?

6) Why are rats a good animal to use when studying human anatomy?

7) Humans are bipedal apes from the order Primates. Primates came from a tree shrew, which is a rodent. All mammals are derived from rodents. We share 90% of our genetic material with rodents; it is just in a different order. Why is it useful to study rats when doing medical experiments for human health?

8) How do the similarities in the anatomy show common descent between humans and rodents?

9) How does the anatomy of rats and humans show evolution has taken place?

What's in That System?

Directions:

Using what you have learned so far and your **textbook** or **internet**, give a general description of what each body system does, list each of the organs in that system, and tell their functions.

1) Nervous system description:

    a. Organs:

2) Digestive system description:

    a. Organs:

3) Integumentary system description:

    a. Organs:

4) Endocrine system description:

    a. Organs:

5) Excretory system description:

    a. Organs:

6) Skeletal system description:

    a. Organs:

7) Muscular system description:

    a. Organs:

8) Immune system description:

    a. Organs:

9) Circulatory system description:

    a. Organs:

10) Respiratory system description:

    a. Organs:

Body System Interactions

Directions:

Watch people doing each of these actions and tell which and how body systems directly interact with each other for the following actions. Then tell which and how other body systems indirectly interact with those body systems to keep them functioning.

1) Running

2) Eating

3) Breathing

4) Talking

5) Driving

6) Playing basketball

7) What do your heart and lungs do when you do a lot of exercise?

   a. Why does this happen?

8) When your body gets cold, what does it do?

   a. Why do you think it does this?

9) What does your body do when it gets too hot?

   a. Why do you think this happens?

Baby Doll Project

Directions:

You will need a large **plastic baby doll** (thick enough to hold the other body systems), many colors of **Play-doh** (dries out but does not stain) or **oil-based clay** (does not dry out but does stain), **red**, **black**, and **yellow yarn** and **thread**, and **white pipe cleaners. Looking at the materials and lab we will be using, what are the safety precautions we should take to protect ourselves and materials during the investigation?**

1) While studying human anatomy, I like to have a long project where the students build a baby from the inside out as we study the body systems. As we study each system, the students build that body system on their doll using either Play-doh or oil-based clay. We start with the integumentary system by cutting the doll in half, separating the ventral and dorsal sides. The hollow doll serves as the integumentary system. Each body system will be built inside the doll and have its own color. The order of study and materials of these systems are:

    a. Integumentary – doll
    b. Nervous – (yellow) clay, yarn, and thread
    c. Endocrine – clay
    d. Skeletal – (white) pipe cleaners and clay
    e. Excretory – clay
    f. Digestive – clay
    g. Respiratory – clay
    h. Cardiovascular (red) clay, yarn, and thread
    i. Lymphatic – (black) clay and thread
    j. Muscular – clay

2) Encourage your students to use their imagination and other materials to construct all the parts in the most detail they can.

3) When finished, have students dissect and label their dolls.

Construction of a Human Body Model and Anatomy Book

Introduction:

There is a big market for new students wanting to study in the medical field. Universities are trying to find new ways to educate students. A medical science education company, _____ Education Medical (teacher), wants to hire a firm (your table group), to build a prototype human model out of clay. Your firm will compete with other firms (other table groups) for their business. This model will need to be anatomically correct. It will not have reproduction organs to make it non-gender specific. The model needs to fit in a 2.2-liter airtight rectangle container. Undergraduate medical students would carry the model around and use it as a reference tool to learn all the human body parts in class and at home. If your design wins, you could earn millions of dollars when your models are sold to universities worldwide. _____ EM wants to make sure that the model is color-coded for each organ system. You will be given a competitor's model to look at (The Invisible Man). We want you to study it and make improvements. This model is missing the endocrine, muscular, and skeletal systems. We want those systems included in your model. See what you can learn from it to make your model. The neat thing about your model is that someone will be able to dissect it when done. You may want to show off this feature when the representatives from _____ EM look at your model.

This model will include a book that shows all the organ systems. Every two pages in the book will show an organ system. One page will have a picture of that system labeled; this should show where the organs are located in relation to the other organs in that system. The other page will tell how each of the organs function. We have some anatomy books available for you to look at to see how yours should be formatted. Having pictures of your model showing the labeled organs might impress _____ EM. Make sure to mention the dissection feature of the model in the book. That technology has never been developed until now; it could be a good selling point.

Materials you need to get for your group:

-Invisible Man doll

-Tube Party Pack of 10 mini cans of Play-doh compound (sold at Wal-Mart and Target)

- An airtight sealed container to store the model in so it does not dry out (Rubbermaid, Ziploc, or Glad 2.2-liter rectangle containers)

Filling job descriptions:

Your firm needs to fill the jobs that have opened up for this project. One type of job is the Human Anatomy Sculptor. The other type of job is Author for the Human Anatomy Book, which goes with the model. Let us look at the job duties below to see which type of job each team member will fill. There will be two students to fill each job. Each member needs to contribute to at least four organ systems in the project, stay on task to keep pace with their group and the class and communicate at least four times a day with a team member in a positive way to have their part of the project fit with the rest and share in clean up responsibilities.

Duties of the Sculptor:

The Sculptor will form the body by putting together the organ systems using Play-doh. They need to take a picture of each organ system as they complete it and send it electronically to the author. You will have two sculptors, each working on different organ systems in a way, so they fit in the same model.

Duties of the Author:

Your group will take electronic pictures to use as diagrams in the book. The authors are to list each organ system's organs and define their function using Word or PowerPoint. When they receive the pictures electronically, they need to label them using the same software to keep the style and font the same for all body systems.

How this will look:

You will evaluate and be evaluated by all your group members in the next page's rubric categories. We need to understand what each behavior and category looks like on the rubric. Your teacher will ask volunteers to show/tell what they think a 2, 1, and 0 look like for each behavior. After students act it out or describe each category for the behavior, your teacher will give a correction to let them know if they were accurate or not.

Cooperative Task Rubric

Points	2	1	0
Final Project	Contributed to at least 4 organ systems in the project	Contributed to 2-3 organ systems in the project	Contributed to less than 2 organ systems in the project
Stayed on task	Kept pace with the group and class every day	Kept pace with group and class most days	Did not keep pace with the group and class most days
Amount of communication	Communicated at least 4 times a day with a member of their group to make sure their part of the project fit with the rest	Communicated 1-3 times a day with a member of their group to make sure their part of the project fit with the rest	Didn't communicate with members of their group to have their part of the project fit with the rest
Positive and negative communication	Communication was always positive	Communication was positive most of the time but not all of the time	Communication was negative most of the time
Clean up	Helped clean up every day	Helped clean up most days	Did not help clean up most days

Timeline and the order in which you will build your project:

Today is the first day of the project. We will learn some background information about what we will do and organize your group. We also see how this information is used in real life. Next, you will work on each body system after gathering information and doing labs on those body systems. To do this, we will build organ systems one at a time, starting with the Nervous System and finishing up with the Integumentary System. You will then present your project for an evaluation, evaluate other projects, and evaluate your own.

How to build the model:

Use your textbook, class anatomy books, and the model provided to help you sculpt each organ system. Your teacher will have examples of the two systems we will focus on each day. We will build each organ system with Play-doh in the order shown below (Starting on the inside and finishing on the outside).

1) Nervous system
2) Endocrine system
3) Skeletal system
4) Excretory system
5) Digestive system
6) Respiratory system
7) Cardiovascular system
8) Muscular system
9) Integumentary system

How to build your book:

We have gone over all the body systems earlier in class. You can use information in your notes, textbook, activities, and class anatomy books to build your book. As your teammates build each organ system, you can take pictures of your model with your cell phone and send the pictures to one of your author's email. The author can then open their email and copy and paste the pictures to Word or PowerPoint. They can then manipulate the pictures to label them on one of the class computers to build the book. If you do not have a cell phone, you can use the class digital camera to take pictures and download them to a computer in the room to build your book. If this is too technical for your group, you may want to draw and label your pictures of the organ systems with colored pencils. This procedure will be a good option if you have someone in your group who is a good artist. Many medical diagrams have been drawn in textbooks and reference books this way. Make sure your book looks professional.

How you will be graded:

An agency hired by _____ EM (student groups from another class) will rate the models and manuals on a scale of 1-10 to select the winners for your class. After you evaluate other groups' products, you will rate your own. You will be graded on how accurately you placed organs in each organ system. Someone must be able to recognize the shape of the organ, and it needs to be labeled with the correct name. The next thing your book will be graded on is how often you used proper labeling techniques in each organ system. Your labeled pictures must not cross label lines, and words must all be horizontal. Your book will also be graded on how accurately you describe each organ's functions in each organ system. Lastly, your book will be graded on its professional appearance. There are 4 points possible in each category for each organ system. If you leave out an organ system, you will receive no points for all four categories in that system; so you will lose 16 points for each organ system you leave out of your book (4 x 4 = 16). There are 16 points possible for each of the nine organ systems totaling 144 points possible for the book (16 x 9 = 144). There is a rubric below and a score sheet on page 330 to help you see how you will be graded for your sculpture and Anatomy Book using the rubric.

Rubric for Anatomy Book (144 points are possible):

Points per Organ System for each Category	4 points	3 points	2 points	1 point	0 points
Correct Placement of Organs	All organs are in the correct place	1 error in placement of organs	2-3 organs have placement errors	> 3 and < half the organs have placement errors	> half the organs have placement errors
Labeling Techniques	No label lines crossed, and all wording is horizontal	1 label line crossed or words not horizontal	2-3 label lines crossed or words not horizontal	> 3 and < half the label lines crossed or words not horizontal	> half the label lines are crossed or words not horizontal
Organ Function Accuracy	Functions of all organs defined correctly	1 error in defining the functions of the organs	2-3 errors in defining the functions of the organs	3-5 errors in defining the functions of the organs	> 5 errors in defining the functions of the organs
Professional Appearance	The picture is clearly identifiable and precise, and all writing is typed	The picture is reasonably identifiable but not precise, and all writing is typed	The picture is reasonably identifiable but not precise, and some or all writing is neatly printed (not typed)	The picture is not identifiable, or writings are not neatly printed (not typed)	The picture is not identifiable, and writings are not neatly printed (not typed)

Score Sheet for Anatomy Book (teacher will use)

	Placement of Organs	Labeling Techniques	Organ Function Accuracy	Professional Appearance	Points Earned for Each System
Nervous system					
Endocrine system					
Skeletal system					
Excretory system					
Digestive system					
Respiratory system					
Cardiovascular system					
Muscular system					
Integumentary system					
Points Earned in Each Category					Total Points _____

Virtual Investigations that go with Animal System Interactions

ExploreLearning.com

Homeostasis Gizmo

Human Homeostasis Gizmo

Disease Spread Gizmo

Hearing: Frequency and Volume Gizmo

Sight vs. Sound Reactions Gizmo

Circulatory System Gizmo

Digestive System Gizmo

Senses Gizmo

Muscles and Bones Gizmo

Homeostasis STEM Case Gizmo

Homeostasis Handbook Gizmo

PhET.colorado.edu:

Color Vision

Eating and Exercise

Neuron

Simplified MRI

Sound

Biology TEKS and NGSS Correlations:

Nature of Science Concept Map Bio c 1ABH 2AD 3AB 4AB

Focus on the Process Bio c 1ABCDEFG 2AC 3ABC 4A

Measurement Lab Bio c 1ABCDEF 2BC 3AB 4A

Patterns in Pennies Bio c 1ABCDEF 2BCD 3ABC 4A

Virtual Introduction Investigations Bio c 1ABEFG 2ABCD 3ABC 4AB

Nature of Life Concept Map Bio c 5ACD 7A; HS-LS1-16

Scale Model of a Hydrogen Atom Bio c 1ABCDEG 2A 3ABC 4AB; HS-LS1-6

Building Bohr Models Bio c 1ABCDEFG 2ABC 3ABC 4AB; HS-LS1-6

Models of Micro-molecules Bio c 1ABCDEFG 2ABC 3AB 4A 5C 13C; HS-LS1-6

Models of Maco-molecules Bio c 1ABCDEFGH 2ABC 3ABC 4AC 5AC 7AC 9A 13C; HS-LS1-6

DNA Extraction Bio c 1ABCD 2AD 3AB 4A 5A 6A 7AD 11B 12A; HS-LS1-16

Practice Reading Nutrition Labels Bio c 1BE 2B 3B 4ABC 5AC 13C; HS-LS1-6

Model of Denaturing an Enzyme Bio c 1ABCDEFG 2ABC 3AB 4A 5A 11B; HS-LS1-6

Virtual Characteristics of Life Investigations Bio c 1 ABCDEG 2ABCD 3AB 4A 5A 7A 11B; HS-LS1-16

Ecology Concept Map Bio c 13ABCD; HS-LS2-1245678

Environmental Issues and Ethics Bio c 1ABEF 3ABC 4AC 13ABCD; HS-LS2-7

Hierarchical Organization of Ecosystems Bio c 1ABEFG 2A 3AB 4A 13ABC; HS-LS2-4

Competitive Relationships Bio c 1AB 3AB 4A 13ABC; HS-LS2-18

Symbiotic Relationships Bio c 1AB 3AB 4A 13ABC; HS-LS2-8

Build an Ecosystem Bio c 1ABCDEG 2ABC 3ABC 4A 11A 13A; HS-LS2-4

Ecological Pyramid Bio c 1CDEG 2ABC 11A 13A; HS-LS2-4

Social Behaviors Bio c 1ABE 3AB 4A 13A; HS-LS2-8

What Happens to the Food Web? Bio c 1ABEG 3AB 4A 13ABD; HS-LS2-12567

Making a Food Web Bio c 1CF 2AB 11A 13ABD; HS-LS2-12567

Population Count Bio c 1ABCEG 2ABCD 3ABC 4A 13AD; HS-LS2-2

Population Growth Curves Bio c 1ABFG 2B 3AB 4A 13AB; HS-LS2-12

Causes for Invasive Species Bio c 3AB 13ABD; HS-LS2-7

Humans Changing Ecosystems Bio c 1ABE 3AB 4A 13ABD; HS-LS2-7

Ecosystem Research Report Bio c 1ABF 3B 4C 11A 13ABD; HS-LS2-45

Biomes Chart Bio c 1ABF 3B 4C 13ABC; HS-LS2-45

Composition of the Atmosphere Bio c 1B 4BC 9B

The Greenhouse Effect Bio c 1ABCDEFGH 2ABCD 3ABC 4AB 13D

Climate and Greenhouse Gases: Data Table Bio c 1ABFH 2B 3ABC 4A 13BCD

Carbon Dioxide and Population Bio c 1ABF 2B 3ABC 4AB 13D

Climate Change Bio c 1AB 3ABC 4ABC 13ABCD

How Life is Allowed on Earth? Bio c 1AB 3ABC 4AB 13BCD; HS-LS1-1567

Our Little Mountain Bio c 1BF 2B 13BD; HS-LS2-6

Primary or Secondary Succession Bio c 1ABCE 3AB 4A 13BD; HS-LS2-6

Natural and Manmade Disasters Bio c 1B 3B 4AB 13BCD; HS-LS2-67

Virtual Ecology Investigations Bio c 1ABEFG 2ABC 3ABC 4AB 13ABCD; HS-LS2-1245678

Evolution Concept Map Bio c 9AB 10ABCD; HS-LS4-126

Analyzing an Ancient Puzzle Bio c 1AB 2AB 3ABC 4AB 9A; HS-LS4-1

The Story of Life Bio c 1BFGH 2ABC 3ABC 4ABC 9AB 10CD 11A 12A 13BCD; HS-LS4-15

Fossil Evidence of Relative Dating Bio c 1ABEG 2B 3AB 4A 9AB; HS-LS4-1

Nuclear Decay Half-life of Pennies Bio c 1ABCDEFG 2ABCD 3AB 4A 9A; HS-LS4-1

Homologous Structures Bio c 1ABEH 2AB 3ABC 4AB 9AB; HS-LS4-1

Evidence in Embryonic Development Bio c 1ABH 2A 3ABC 4A 9A; HS-LS4-1

Comparing Relatedness with Proteins Bio c 1ABEFGH 2ABCD 3ABC 4AB 9AB 10C; HS-LS4-1

Biodiversity in Ecosystems Bio c 1ABEF 3AB 4AB 10BC; HS-LS2-2

Changing Environments for Beads Bio c 1ABCDEFG 2ABD 3AB 4A 10BC; HS-LS4-234

Variation Within a Population Bio c 1ABCDEFGH 2ABCD 3ABC 4AB 10ABC; HS-LS4-234

Goldfish Evolution Bio c 1ABCEFG 2ABCD 3ABC 4AB 10ABCD 13A; HS-LS4-234

Domains Bio c 1AB 3AB 4A 9A

Functions the Kingdoms Serve Bio c 1AB 3AB 4A 9A

Relationships in Classification Bio c 1ABFG 2AB 3AB 4A 9A

Pamishan Dichotomous Key Bio c 1ABE 3B 9A

Classifying Animals Bio c 1ABEF 2B 3BC 4AB 9A 13D

Virtual Evolution Investigations Bio c 1ABEFGH 2ABCD 3ABC 4AB 9AB 10ABCD; HS-LS4-123456

Prokaryotic Life Concept Maps Bio c 5ABC 13A; HS-LS1-34, 2-3

Effects of Diffusion on Cells Bio c 1ABCDF 3AB 4AB 5BC 11A 12A; HS-LS1-3

Osmosis Toothpicks Bio c 1ABCDG 3AB 4A 11A 12AB; HS-LS1-3

Membrane Models Bio c 1ABCDEG 2AB 3AB 4A 5AC 12A

Seeing Different Types of Bacteria Bio c 1ABCDF 3AB 4A 5B

Seeing Live Bacteria in Yogurt Bio c 1ABCDF 3AB 4A 5B

Draw a Detailed Picture of Bacteria Bio c 1BF 5AB

DNA Replication Bio c 1ABFG 2ABC 5A 6A 7A; HS-LS1-4

Virtual Prokaryotic Life Investigations Bio c 1ABEFG 2ABC 3ABC 4AB 5ABC 6A 7A; HS-LS1-34

Unicellular Protists Concept Maps Bio c 5ABCD 6ABC; HS-LS1-4

Cell Town Bio c 1ABFG 2AB 5ABC 11AB 12A

Characteristics of Prokaryotic and Eukaryotic Cells Bio c 1ABCDEFH 3ABC 4AB 5AB 9A

Can You Make the Connection? Bio c 1ABCEFG 2AB 3AB 4AC 5ABD

Seeing Cell Division Bio c 1ABCDEF 3AB 4A 5ABC 6AC; HS-LS1-4

Hand Models Showing Cell Division Bio c 1ABCG 2AD 3AB 4A 6A; HS-LS1-4

Cell Cycle Timing (With Slides) Bio c 1ABCDEF 2BD 3ABC 4A 5A 6A; HS-LS1-4

Cell Cycle Timing (With Pictures) Bio c 1ABEFG 2ABC 3ABC 4A 6A; HS-LS1-4

Virtual Unicellular Eukaryotes Investigations Bio c 1ABEFG 2ABC 3AB 4AB 5C 6AC; HS-LS1-4

Genetics Concept Maps Bio c 6ABC 7C 8AB 10C; HS-LS3-123

Modeling Meiosis Bio c 1BCDEFG 2ABC 3A 4A 6A 8A 10D; HS-LS3-123

Comparing Ratios for Monohybrid Cross Bio c 1ABCDEFGH 2ABCD 3AB 8AB; HS-LS3-3

Paper Mates Bio c 1BCDEFG 2ABC 3AB 8AB; HS-LS3-123

Create a Baby Bio c 1ABCDEFG 2ABC 3AB 4A 8AB 10D; HS-LS3-123

Construct Your Family Pedigree Bio c 1BEFG 8AB; HS-LS3-1

Virtual Genetics Investigations Bio c 1ABEFG 2ABCD 3AB 4AB 6A 7C 8AB; HS-LS3-123

Gene Expression Concept Map Bio c 5D 6BC 7BCD; HS-LS1-1

Protein Synthesis of the Quaddie Bio c 1ABEFG 2AB 3ABC 4A 5A 7AB; HS-LS1-1

Protein Synthesis Role Play Bio c 1ABG 3AB 4A 5A 6AC 7ABC 10D; HS-LS1-16

Making a Karyotype Bio c 1ABCDEFG 2ABCD 3AB 4AB 6C 7CD 8A; HS-LS3-12

DNA Fingerprinting Bio c 1ABEG 2AB 3AB 4A 7D 8AB; HS-LS3-1

Lego Mutation Models Bio c 1ABCDEG 3AB 4A 7C 8A; HS-LS3-2

Virus Lytic Cycle Bio c 1ABEFG 2B 3AB 4AC 5D; HS-LS3-2

Virtual Gene Expression Investigations Bio c 1ABEFG 2ABCD 3ABC 4AB 5C 6BC; HS-LS1-1, 3-2

Plants Concept Map Bio c 5AC 6B 9A 11AB 12AB; HS-LS1-2367, 2-3

Evolution of Plants Bio c 1ABEFGH 2AB 3AB 4AC 5C 6B 9AB 10C 11AB 12AB; HS-LS4-5

Root, Stem, and Leaf Cross Sections Bio c 1ABCDF 3AB 4A 5AC 6B 11A 12AB; HS-LS1-2

Tap Roots vs. Fibrous Roots Bio c 1ABEF 3AB 4A 5C 6B 11A 12AB; HS-LS1-2

Monocot vs. Dicot Bio c 1ABCE 3A 4AB 11A 12AB; HS-LS1-2

Germination and Topisms Lab Bio c 1ABCDEF 2BC 3ABC 4A 5C 6B 11AB 12AB; HS-LS1-3

Conservation of Life: Photosynthesis and Respiration Bio c 1ABEG 3AB 4A 5A 11A; HS-LS1-567, 2-3

Building a Model of a Water Molecule Bio c 1ABG 2AD 3AB 4A 5C 12B; HS-LS1-3

Celery Transport Bio c 1ABCDGH 2AB 3ABC 4AB 5C 11A 12AB; HS-LS1-3

Transpiration Pull Bio c 1ABCDE 2AB 3ABC 4AB 5C 11A 12AB; HS-LS1-3

Seeing a Stoma Bio c 1ABCDEF 2AB 3AB 4A 5C 11A 12AB; HS-LS1-3

Flower Dissection Bio c 1ABCDEFG 2AB 3AB 4AB 12AB; HS-LS1-2

Seed Dissection Bio c 1ABCDEFG 3AB 4A 5AC 6B 11A 12AB; HS-LS1-2

Virtual Plants Investigations Bio c 1ABEFG 2ABCD 3ABC 4AB 5AC 6B 11AB 12AB; HS-LS1-23567, 2-3

Evolution of Animals Concept Map Bio c 5C 9A 10C 12AB; HS-LS4-5

Evolutionary Relationships Seen Through... Bio c 1ABEFGH 2ABC 3ABC 4A 5A 7ABC 9A 10CD; HS-LS4-15

Comparing Vertebrates Bio c 1ABEFG 2ABC 3ABC 4A 5A 7BC 9A 10CD; HS-LS4-15

Hydra Lab Bioc c 1ABCDEFG 3AB 4A; HS-LS1-23, 2-3

Observing Flatworms and Roundworms Bio c 1ABCDEFG 3AB 4A 12AB; HS-LS1-2

Planaria Lab Bio c 1ABCDEFG 3AB 4A 12AB; HS-LS1-23, 2-3

Earthworm Dissection Bio c 1ABCDEFG 3AB 4A 12AB; HS-LS1-2, 2-3

Crawfish Dissection Bio c 1ABCDEFG 3AB 4A 12AB; HS-LS1-2, 2-3

Virtual Evolution of Animals Investigations Bio c 1ABEFG 2ABC 3ABC 4A 6B 9A 10C; HS-LS1-2, 4-15

Animal System Interactions Concept Map Bio c 11B 12AB 13A; HS-LS1-2, 2-3

Recognizing Body Tissues Bio c 1BCDEFG 2AB 3AB 5C 6B 12AB; HS-LS1-2

Heart Dissection Bio c 1ABCDEFG 3AB 4A 5C 11A 12AB; HS-LS1-2, 2-3

Making a Working Lung Bio c 1ABCDG 2ABD 3AB 4A 5C 11A 12AB; HS-LS1-2, 2-3

Measuring Respiration of an Animal Bio c 1ABCDE 2BC 3AB 4A 5C 11A; HS-LS1-3, 2-3

Control of Human Respiration Bio c 1ABCDEFG 2ABC 3AB 4A 5C 11A 12AB; HS-LS1-3, 2-3

Human Respiration Bio c 1ABCDEFG 2ABCD 3AB 4A 5C 11A 12AB; HS-LS1-3, 2-3

Measuring Heart rate and Physical Fitness Bio c 1ABCDE 2BC 3AB 4A 5C 11A 12AB; HS-LS1-3, 2-3

Observing Vertebrate Skeletons Bio c 1ABCDEF 2B 3AB 4A 9A 12A; HS-LS1-2

Brain Dissection Bio c 1ABCDEFG 2B 3AB 4A 9A 12AB; HS-LS1-2

Eye Dissection Bio c 1ABCDE 2AB 3AB 4A 12AB; HS-LS1-2

Kidney Dissection Bio c 1ABCDEFG 2AB 3A 4A 12AB; HS-LS1-2

Ain't Nothing but a Chicken Wing Bio c 1ABCDEFG 2B 3AB 4A 5AC 9A 11A 12AB; HS-LS1-2

Fish Dissection Bio c 1ABCD 2B 3AB 4A 5C 9A 11A 12AB; HS-LS1-2, 2-3

Dissection Comparing a Rat to a Human Bio c 1ABCDEF 2BC 3AB 4A 5C 9A 11AB 12AB; HS-LS1-2, 2-3

What's in that System? Bio c 1ABE 2AB 3AB 4AC 5C 11AB 12AB; HS-LS1-2, 2-3

Body System Interactions Bio c 1ABE 3AB 4A 5C 11AB 12AB; HS-LS1-2, 2-3

Baby Doll Project Bio c 1ABCDEFG 2AB 5C 11AB 12AB; HS-LS1-2

Constructing a Human Body Model… Bio c 1ABCDEFG 2ABCD 3AB 4ABC 5C 11AB 12AB; HS-LS1-2

Virtual Animal System Interactions Investigations Bio c 1ABEFG 2ABCD 3ABC 4AB 5C 11AB 12AB; HS-LS1-23, 2-3

Equipment List for all Investigations

If you want to be able to do all the labs in this manual, here is the list of all the equipment you will need in order of appearance:

Small Lego sets

Scales

Meter sticks

Temperature probes

Interfaces

Computers

Logger Pro software

Rulers

Pennies

Roll of pennies

Empty penny rolls

Golf balls

Beads (three colors)

Film cases

Molecular model kit

Periodic Table

Nutrition labels

Stopwatches

Scotch tape

Scissors

Ziploc bag

Textbook

Safety goggles

Play-doh

Butcher paper

Glue

Colored pencils

Ring stands and clamps

Glass beakers of different sizes

Tap water

Shoebox sized plastic tubs colored black on the inside

Press'n Seal Wrap

Paper towels

Salt

Graduated cylinders

Leaves

Shelled peanuts

Yellow Cheese flavored Goldfish crackers

Pretzel Goldfish crackers

Anacharis elodea

Mega Blocks

Compound light microscopes

Lens wipes

Slides and coverslips

Dawn dishwashing soap

Strainers or colanders

Marbles

Dry beans

Buckets

Shoebox sized plastic tubs

Prepared slides of round, rod, and spiral-shaped bacteria

Prepared slides of amoeba

Prepared slides of euglena

Prepared slides of paramecium

Plain or vanilla yogurt

Toothpicks

Plain M&M's

Peanut M&M's

Prepared slides of onion root tips

Prepared slides of whitefish blastula

Different colored pipe cleaners

Twisted phone cords

Masking tape

Connecting alphabet baby letters

Prepared slides of a cross-section of roots monocot

Prepared slides of a cross-section of roots dicot

Prepared slides of cross-sections of stems monocot

Prepared slides of cross-sections of stems dicot

Fresh carrots, celery, lettuce, and onions (used for chives)

Monocot flowers

Dicot flowers

Ziploc freezer bags

Staplers

Pressure sensors and tube setup

Scalpels

Dissecting probes

Dissecting scissors

Dissecting pins

Dissecting pans with rubber bottom

Shelled sunflower seeds

Prepared slides of flatworms

Prepared slides of roundworms

Live cultures of planaria

Pond water

Watch glasses

Dissecting microscopes

Live cultures of hydra

Live cultures of Daphnia

Preserved earthworms

Hand lenses

Preserved crawfishes

Small bags

Respiration monitor belts

Hand-grip heart rate monitors

Models of different animal skeletons

An assortment of animal bones

Model of brain

Preserved sheep brains

Cow eyes

Model of eye

Round balloons

Bread bags

Oxygen gas sensors

Carbon dioxide gas sensors

Kidney model

Preserved sheep kidneys

Chicken wings

Prepared slides of skeletal, smooth, and cardiac muscle

Safety goggles

Preserved perch

Forceps

Preserved rats

String

Model of a human torso

Heart model

Preserved sheep hearts

Model of lungs

Model of working lungs

Oil-based clay

Red, black, and yellow yarn

Red, black, and yellow thread

White pipe cleaners

Invisible Man doll

2.2 liter airtight sealed container